Fiber Optics - From Fundamentals to Industrial Applications

Edited by Patrick Steglich and Fabio De Matteis

Published in London, United Kingdom

IntechOpen

Supporting open minds since 2005

Fiber Optics - From Fundamentals to Industrial Applications
http://dx.doi.org/10.5772/intechopen.74877
Edited by Patrick Steglich and Fabio De Matteis

Contributors
Yao Zhang, Xiaoshuai Liu, Yoshito Shuto, Jing Wang, Zhensheng Jia, Alberto Campos, Curtis Knittle, Monika Bahl, Patrick Steglich, Fabio De Matteis

First published in London, United Kingdom, 2019 by IntechOpen
IntechOpen is the global imprint of INTECHOPEN LIMITED, registered in England and Wales, registration number: 11086078, The Shard, 25th floor, 32 London Bridge Street
London, SE19SG - United Kingdom
Printed in Croatia

British Library Cataloguing-in-Publication Data
A catalogue record for this book is available from the British Library

Additional hard and PDF copies can be obtained from orders@intechopen.com

Fiber Optics - From Fundamentals to Industrial Applications
Edited by Patrick Steglich and Fabio De Matteis
p. cm.
Print ISBN 978-1-83881-155-6
Online ISBN 978-1-83881-156-3
eBook (PDF) ISBN 978-1-83881-157-0

For EU product safety concerns:
IN TECH d.o.o., Prolaz Marije Krucifikse Kozulić 3, 51000 Rijeka, Croatia, info@intechopen.com or visit our website at intechopen.com.

Meet the editors

Patrick Steglich is a research associate at the IHP—Leibniz-Institut für innovative Mikroelektronik, Germany, and lecturer for photonics and optical technologies at the Technical University of Applied Sciences Wildau, Germany. He has held a Master's degree in Photonics from the Technical University of Applied Sciences Wildau since 2013. In 2017, he received his PhD in Industrial Engineering from the Università degli Studi di Roma "Tor Vergata" for his work in the field of integrated photonics for communication and sensing. His research is focused on emerging photonic devices and waveguide concepts for telecommunication and sensing applications.

Fabio De Matteis graduated in Physics at the University of Rome-LaSapienza in 1987 and received a PhD in Physics from the University of Antwerp, Belgium, in 1993. He has been at the University of Rome "Tor Vergata" since 1995.
His scientific interest has been devoted to: the synthesis and characterization of materials for photonics by means of spectroscopic techniques (absorption and luminescence, static and time resolved by ultrafast techniques, Raman); synthesis and nonlinear optical characterization of materials for optoelectronics; polar ordering of nonlinear chromophores induced by corona and electrode poling; nanostructuring (mono- and bidimensional gratings) in insulating materials by photolithographic techniques; biopolymeric scaffolds by two-photon polymerization for tissue engineering; and impedance spectroscopy of solid electrolyte materials showing ionic conduction.

Contents

Preface

Optical fibers in metrology, telecommunications, sensors, manufacturing, and health science have gained massive research interest. This book covers materials, components, and systems across the application spectrum, from interferometers for spectroscopy to fiber optic applications in medicine. Furthermore, the number of fields of application of optical fibers is increasing at a fast pace. On the other hand, the literature on optical fibers and the foundations of their functioning are extremely vast and we have no ambition to make an original contribution in this respect. This book aims to present a collection of recent advances in fiber optics. It covers the current progress and latest breakthroughs in emergent applications of fiber optics and specialty fibers.

The book includes five contributions on recent developments in optical fiber communications and fiber sensors, as well as the design, simulation, and fabrication of novel fiber concepts.

The editors Patrick Steglich and Fabio De Matteis provide a brief historical overview of important developments in the field of optical fibers, as well as current trends and perspectives.

Xiaoshuai Liu and Yao Zhang review the multifunctional manipulation for biological cells based on the elaborately designed fiber probes.

Monika Bahl describes the propagation of structured light fields in optical fibers, which find important applications in ICT but also in the design of ultrapowerful microscopes and novel spectroscopic analyses.

Yoshito Shuto presents a theoretical study on the evolution of a fiber fuse in a single-mode optical fiber. A novel nonlinear oscillation model using the Van der Pol equation is described and both the silica-glass densification and cavity formation observed in fiber fuse propagation are qualitatively explained.

Jing Wang, Zhensheng Jia, Luis Alberto Campos, and Curtis Knittle propose and demonstrate a delta-sigma digitization and coherent transmission of DOCSIS 3.1 signals for the first time.

We thank all authors for their contributions.

Finally, we express a great deal of thanks to the editing staff of IntechOpen, particularly Ms. Lada Bozic, for all their efforts.

Dr. Patrick Steglich
IHP—Leibniz-Institute for Innovative Microelectronics,
Germany

Technical University of Applied Sciences Wildau,
Germany

Dr. Fabio De Matteis
University of Rome "Tor Vergata",
Italy

Section 1

Introduction

Chapter 1

Introductory Chapter: Fiber Optics

Patrick Steglich and Fabio De Matteis

1. Introduction

Whether covering a few meters or hundreds of kilometers, optical fibers allow to transmit information in a short time. Without communication over long distances, humankind would not have got far. It is the basis of every economic and political development. Up to the twentieth century, however, it was still the stagecoach that brought the information from A to B most quickly—at a speed of no more than 40 km an hour. Nowadays, optical fibers and light-wave technologies enable high-speed data communication.

In summary, this means that long-distance communication works in principle just like a Morse telegraph: a combination of pulsed signal travels in a medium bringing a coded information. The only difference is that we no longer send the signals with electricity, but with light. Indeed, optical fibers carry light in the visible and near-infrared region (~100 THz), and therefore, they are often called light-wave systems to distinguish them from microwave systems (~1 GHz) [1].

The principle is actually the same as in the Morse telegraph. Just as at the end of the 18th century, information is assigned to specific signals or light pulses. However, the copper cable was unsuitable for this purpose. In 1965, Manfred Börner, a German physicist, used a laser instead of a broadband light source to transmit information with optical fibers for the first time.

A typical fiber optic cable, as used by Manfred Börner and currently used all over the world, consists usually of a fiberglass core covered by a fiberglass cladding. To protect against external influences and to prevent scratches or dirt and moisture from penetrating, the entire fiber is covered with a protective layer. The material of the cladding must have a smaller refractive index than the material of the core [2]. This allows for total internal reflection, a phenomenon well known since the first half of the nineteenth century (Colladon and Babinet, Tyndall).

It is noteworthy that for more than a century the total internal reflection was mostly used for illumination systems and image transmission to short distances. The first person to try to transmit images by means of a bundle of optical fibers was Heinrich Lamm, the first to propose a fiber-optic endoscope in 1930 [3].

In 1954, van Heel [4], on one side, and Hopkins and Kapany [5], on the other, simultaneously reported a way of conveying optical images along a glass fiber; in particular, the former writes: "*Preliminary experiments, started in January 1950, have shown that coating the fibres with silver or any other metal yields an unsatisfactory transmission. A much better result was obtained when the fibres were coated with a layer of lower refractive index, which ensured total reflexion*". The optical fiber was born.

However, for a breakthrough in optical communications, it has been necessary to wait for Börner to first use a fiber-optic cable in combination with a laser. So here he had built a true optical data transmission system. He also employed photodetectors at the end of the fiber optic cable. In 1966, Börner applied for a patent for the system for the targeted transmission of information via optical fibers for the company AEG-Telefunken. And then, it finally came out of the experimental phase

IntechOpen

into the technical realization—until today. All optical long-distance transmission systems are also currently working according to this system principle proposed by Manfred Börner.

However, there was still a lot to improve. Because the data transfer did not really work well as the light intensity was simply too low, only short transmission distances were possible. One of the solutions to—at least this problem—was actually quite simple. The fiber material had minimal scratches, cracks, and bumps. Charles Kuen Kao has improved the fiber glass in 1966 and solved the problem. He received the Nobel Prize in 2009 for this and for further developments in fiber optic technology.

And yet, even the best and clearest glass fibers have natural limits. Therefore, it is necessary to amplify the light signal after a certain distance. Initially, intermediate stations were set up, in which the incoming light signal was first converted into an electrical signal—and then again into a light signal. As a result, it had the initial light intensity again and could be transmitted over a long distance. However, this solution was unsatisfactory because the intermediate stations drastically increased transmission time.

To solve this problem, purely optical amplifiers have been developed. With this, it was finally possible to guide the light without interruption over very long distances. The most important and widespread optical amplifier is the erbium-doped fiber amplifier [6]. In this case, a part of the cable is made of a special material that amplifies the light synchronously and seamlessly. The erbium-doped fiber amplifier works much like a laser where the active material (erbium), incorporated in the cable, is pumped to a population inversion state producing a regeneration of the signal intensity by stimulated emission [7]. The data transfer keeps hereby finally its fast pace without interruption of the propagation along the fiber cable. That is why it is on the longest transmission lines in the world, on the transatlantic route between Europe and the United States.

In a further step, several carrier wavelengths are transmitted through a single fiber to increase the information capacity. For this purpose, the wavelengths are all first brought together, which is known as multiplexing. At the receiver side, the wavelengths are separated again by demultiplexing. The further development of fiber transmission is not yet finished. Researchers see further potential, for example, with variable structures of the glass fiber.

A new generation of optical fiber has a fiber core, which is interspersed with holes. In this case, the light is confined not only in the core material but also in the holes or, in particular, in a larger hole in the middle of the fiber core. This works like a veritable turbo in the speed of data transmission. After all, light is extremely fast at around 300,000,000 km/s in vacuum. Unfortunately, the fiber material slows down considerably, by around 30% due to the larger refractive index.

A much better material, then, would simply be air, because it simply slows down less than glass or silicon. That is why researchers at the University of Southampton in 2013 developed a hollow core fiber in which the data could be transmitted at a rate of about 99.7% of the speed of light in vacuum [8]. Thus, data can be transmitted even faster or even more data can be transmitted at the same time.

Current research is also focusing on new modulation formats to increase the data rate and, therefore, to unleash the whole potential of optical communication. Such modulation formats employ different physical properties of light such as the amplitude, phase, and wavelength. During the last decades, several modulation formats like phase shift keying (PSK) and amplitude shift keying (ASK) but also advanced modulation formats such as m-th order quadrature amplitude modulation (QAM), binary phase shift keying (BPSK), and m-ASK were realized [9]. To generate such higher modulation formats, novel electrooptical modulators [10–12] are combined with optical fibers, which is known as active optical fibers. These fibers are of special interest for rack-to-rack applications.

Optical fibers play a crucial role in telecommunication. Applications can be found in many areas such as optical fiber lasers [13], optical fiber interferometers [14, 15], optical fiber amplifier [16], and optical fiber sensors [17]. Especially the latter one has widespread applications in detecting magnetic fields [18], humidity [19], temperature [20], or biological molecules [21].

Author details

Patrick Steglich[1,2*] and Fabio De Matteis[3]

1 IHP—Leibniz-Institut für innovative Mikroelektronik, Frankfurt (Oder), Germany

2 University of Applied Sciences Wildau, Wildau, Germany

3 University of Rome 'Tor Vergata', Rome, Italy

*Address all correspondence to: patrick.steglich@th-wildau.de

References

[1] Agrawal GP. Fiber-Optic Communication Systems. Vol. 222. Hoboken, New Jersey: John Wiley & Sons; 2012

[2] Addanki S, Amiri IS, Yupapin P. Review of optical fibers—Introduction and applications in fiber lasers. Results in Physics. 2018;**10**:743-750

[3] Achord JL, Muthusamy VR. "1—The History of Gastrointestinal Endoscopy" Clinical Gastrointestinal Endoscopy. 3rd ed. 2019. pp. 2-11. DOI: 10.1016/B978-0-323-41509-5.00001-3

[4] Van Heel ACS. A new method of transporting optical images without aberrations. Nature. 1954;**173**:39

[5] Hopkins HH, Kapany NS. A flexible fibrescope, using static scanning. Nature. 1954;**173**:39-41

[6] Trinel J-B et al. Latest results and future perspectives on few-mode erbium doped fiber amplifiers. Optical Fiber Technology. 2017;**35**:56-63

[7] Bigot L, Le Cocq G, Quiquempois Y. Few-mode erbium-doped fiber amplifiers: A review. Journal of Lightwave Technology. 2015;**33**(3):588-596

[8] Poletti F et al. Towards high-capacity fibre-optic communications at the speed of light in vacuum. Nature Photonics. 2013;**7**(4):279

[9] Winzer PJ, Essiambre RJ.Advanced optical modulation formats. In: Optical Fiber Telecommunications VB. Academic Press; 2008. pp. 23-93

[10] Zwickel H, Wolf S, Kieninger C, Kutuvantavida Y, Lauermann M, de Keulenaer T, et al. Silicon-organic hybrid (soh) modulators for intensity-modulation/direct-detection links with line rates of up to 120 gbit/s. Optics Express. 2017;**25**(20):23784-23800

[11] Steglich P, Mai C, Peczek A, Korndörfer F, Villringer C, Dietzel B, et al. Quadratic electro-optical silicon-organic hybrid RF modulator in a photonic integrated circuit technology. In: 2018 IEEE International Electron Devices Meeting (IEDM); 2018

[12] Steglich P, Mai C, Villringer C, Pulwer S, Casalboni M, Schrader S, et al. Quadratic electro-optic effect in silicon-organic hybrid slot-waveguides. Optics Letters. 2018;**43**(15):3598-3601

[13] Duval S et al. Femtosecond fiber lasers reach the mid-infrared. Optica. 2015;**2**(7):623-626

[14] Zhou J et al. Intensity modulated refractive index sensor based on optical fiber Michelson interferometer. Sensors and Actuators B: Chemical. 2015;**208**:315-319

[15] Lee BH et al. Interferometric fiber optic sensors. Sensors. 2012;**12**(3):2467-2486

[16] Firstov SV et al. A 23-dB bismuth-doped optical fiber amplifier for a 1700-nm band. Scientific Reports. 2016;**6**:28939

[17] Hernaez M, Zamarreño CR, Melendi-Espina S, Bird LR, Mayes AG, Arregui FJ. Optical fibre sensors using graphene-based materials: A review. Sensors. 2017;**17**:155

[18] Zheng Y et al. Optical fiber magnetic field sensor based on magnetic fluid and microfiber mode interferometer. Optics Communications. 2015;**336**:5-8

[19] Gao R et al. Humidity sensor based on power leakage at resonance wavelengths of a hollow core fiber coated with reduced graphene oxide. Sensors and Actuators B: Chemical. 2016;**222**:618-624

[20] Hernández-Romano I et al. Optical fiber temperature sensor based on a microcavity with polymer overlay. Optics Express. 2016;**24**(5):5654-5661

[21] Ricciardi A et al. Lab-on-fiber technology: A new vision for chemical and biological sensing. Analyst. 2015;**140**(24):8068-8079

Section 2

Optical Fibers in Biophotnics

Chapter 2

Optical Fiber Probe-Based Manipulation of Cells

Xiaoshuai Liu and Yao Zhang

Abstract

In a liquid environment, optical trapping and multifunctional manipulation of biological cells, in a noncontact, noninvasive, and high-precision way, have become one of the research focuses in the field of integrated optics, biophotonics, and clinical medicine. However, it still faces great challenges to perform multifunctional manipulation in very narrow spaces with high flexibility, including stable retaining, controllable deformation, and precise regulation of a cell chain. Therefore, in this chapter, we introduce the multifunctional manipulation for biological cells based on the elaborately designed fiber probes. With the probes, the sequential organization, precise regulation, and bidirectional transportation of the cell chain were performed. We also discuss the potential applications of fiber probes on the endocytosis and exocytosis purpose, which will play an important role in the detection and treatment of complex disease.

Keywords: optofluidic manipulation, fiber probe, cell chain, biophotonics, fiber tweezers

1. Introduction

In the field of endocytosis and exocytosis, a precise manipulation of a cell was required, especially for the nanomedicine injection, intracellular signaling pathway, and pathogenic research, which holds a great potential in the detection and treatment of complex diseases [1]. Thus, increasing attentions have been paid to the dynamic manipulation of cells in a fluid, which has been proved to be crucial in cell growth [2, 3], differentiation [4–6], drug delivery [7–9], mechanical force transduction [10, 11], etc. Up to now, diverse techniques have shown invaluable performance in the multifunctional manipulation of cells, including dielectrophoretic [12, 13], magnetic field [14, 15], and mechanical force [4, 16, 17]. Nevertheless, these strategies face the challenges of potential sample heating and high power consumption and required attachment of electric/magnetic nanoparticles [18, 19]. Besides, the cells were required to be confined on an elaborated substrate, which was usually not reusable in these techniques [20, 21].

Since first introduced by A. Ashkin, optical tweezers have motivated many intriguing advances in interdisciplinary applications, such as micro/nanophotonics, biphotonics, and biomedicine [22–24]. The most commonly used optical manipulation tool is the conventional optical tweezers (COTs), which uses high numerical aperture objectives to focus the free-space laser beam [25, 26]. After the transfer of photon momentum, the cell will suffer from the optical force, which can be divided into optical scattering force and gradient force, with the direction pointed to the optical propagating direction and beam focus, respectively. The magnitude of optical force was ranged from femtonewton to nanonewton, and thus, it was the ideal chose to measure the response of biological and macromolecular system [27]. Compared to other microscopic techniques, for example, electric filed, magnetic field, and acoustic method, optical tweezers have the great potential of high precision, high flexibility, noncontact, and wide manipulation range. However, COTS faces certain challenges, such as bulky structure, limited integration, and diffraction for nanoparticles [28]. Thus, researchers have developed various schemes of optical tweezers to extend the potential application scenario, including holographic optical tweezers (HOTs), surface plasmon-based optical tweezers (SPOTs), and fiber probe-based optical tweezers (FP).

For HOTs, the optical tweezer system was inserted with a diffractive beam splitter. Then the beam wavefront can be further sculptured, which can manipulate multiple cells simultaneously [29–31]. Nevertheless, appropriate algorithms are required to be elaborately designed for achieving a specific pattern of cells. In addition, the complicated optical system was consisted of dichroic mirrors, spatial light modulators, and high-numerical-aperture focusing objectives. The working distance of objectives limits the depth at which cells can be manipulated in the cell suspensions. Besides HOTs, surface plasmon-based optical tweezers (SPOTs) have also been developed to manipulate cells with high trapping stability and retaining ability, especially for the sub-microsized cells and biological molecules [32, 33]. However, once the substrates designed, the cell pattern is fixed without a flexibility to adjust and transport the cell chain dynamically. Moreover, the organization of cell pattern is also limited at a specific depth of cell suspension. Thus, there is a great need of developing a new strategy to perform the multifunctional manipulation of cells.

2. Fiber probe-based optical tweezers

First, we will give a detailed introduction about the fiber probe-based optical tweezers. As for optical tweezers, a focused laser beam was essential for the stable trapping and dynamic manipulation of cells. To realize a focused laser beam, the fiber probe was designed into a tapered tip, for which the laser beam will be focused by the paraboloid end. After one cell approached the beam focus, it will suffer from the optical gradient force and then attracted into the optical axis. After that, it will be pushed away from the probe tip or attracted into the focus, which depends on the struggling of optical scattering force and gradient force along the optical axis direction. After the cell was trapped one by one, a specific organization of cell pattern can be achieved with the fiber probe. Further, the organized cell pattern can be shifted dynamically in the three dimensional direction. The FPs were fabricated by drawing commercial single-mode optical fibers with a flame-heating technique. By adjusting the stretch parameters, various parabolic ends can be achieved to conduct the multifunctional manipulation of the cell. With the high flexibility, ease fabrication, and compact size, fiber probe-based optical tweezers have been widely used for the trap of dielectric microparticles and cells [28], the shift of fluorescent particles [34], and organelles in the cell [35]. Besides, it is free from the limitation on the depth of cell manipulation in the suspensions and does not require any elaborated substrates, providing a flexible platform that can be easily integrated with microfluidics.

In this chapter, we will discuss the multifunctional manipulation of cells with the designed optical fiber probes (FPs). We will show the precise regulation and bidirectional transport of the cell chain with the FPs. Furthermore, it can also conduct the dynamic rotation and deformation of human red blood cells. With the further combination into the microfluidic technique, FPs have enabled the precise control of cell and further applied into the noninvasive analysis for the endocytosis and extrocytosis behaviors.

3. Optical fiber probe-based manipulation of cells

3.1 The regulation of the cell chain

Figure 1 schematically shows the formation and regulation process of an *E. coli* cell chain. After *E. coli* cells trapped one by one, a cell chain consisted of six *E. colis* was organized at the probe tip. By adjusting FP 2 to approach *E. coli* 5, the cell is rotated and orientated toward the axial direction of FP 2. Then, *E. coli* 5 is removed from the chain with the cell contact sequence changed. By shifting FP 2 along the +*x* direction, *E. coli* 5 is put between the *E. coli* 3 and 4 (**Figure 1c**). After turning off the laser in FP 2, *E. coli* 5 is gradually rotated and orientated along the axial

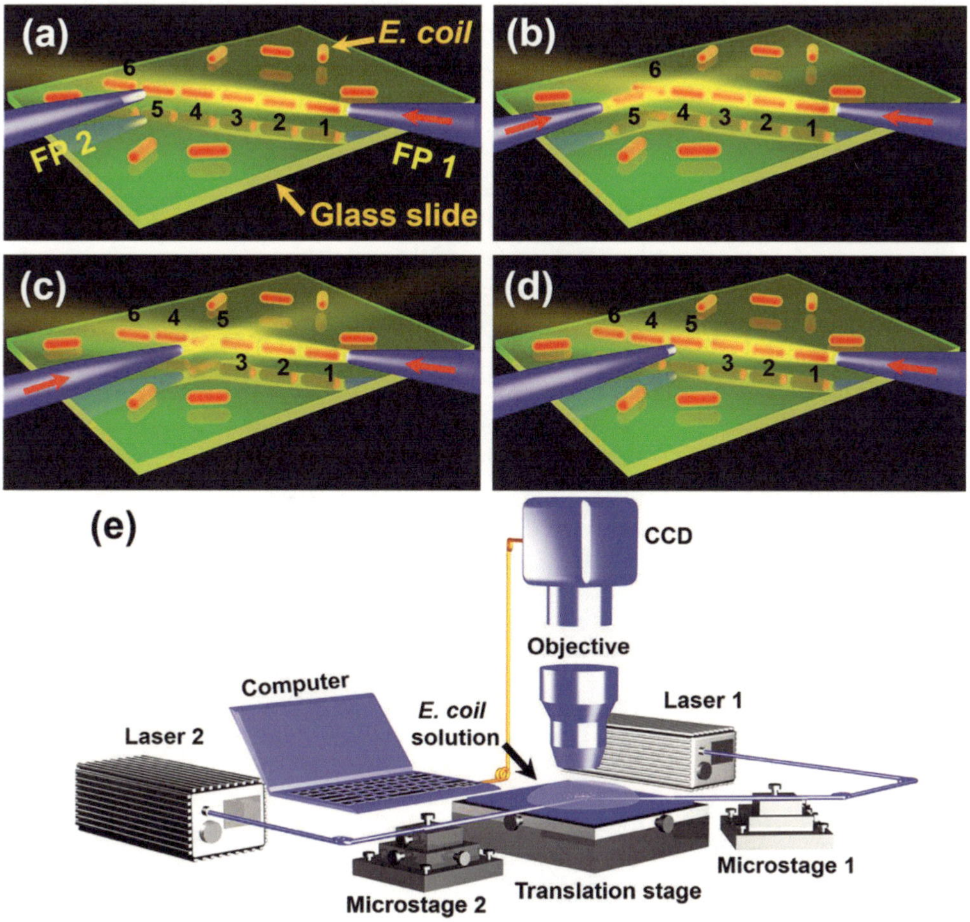

Figure 1.
Schematic of the regulation process and experimental setup. (a) A cell chain is organized at the tip of FP 1. (b) By manipulating FP 2, E. coli 5 is rotated and then removed from the cell chain. (c) E. coli 5 is added back at a new position into the chain. (d) After turning off the laser in FP 2, E. coli 5 is orientated along the axial direction of FP 1. (e) Schematic of the experiment setup [38].

direction of FP 1 (**Figure 1d**), changing the cell contact sequence again. **Figure 1e** schematically shows the experimental setup. After sheathed with a glass capillary, FPs 1 and 2 were fixed on microstages 1 and 2, respectively. The probe tips were immersed in the *E. coli* solution, while their ends were connected to the lasers 1 and 2, respectively. The wavelength of both lasers is set at 980 nm due to the low absorption for the biological cells [36]. Such a wavelength induces little optical damage to bacteria and mammalian cells [37]. The *E. coil* solution was dropped on a glass slide with an injector. The slide was mounted on an *x*–*y* manual translation stage (resolution: 50 nm) to achieve fine positioning and mechanical stability. An optical microscope incorporated with a charged coupled device (CCD) was used for the real time monitoring, image capturing, and video recording.

To demonstrate the operation mechanism, a dynamic regulation of the cell chain was experimentally conducted. After the laser beam injected into FP 1 (P_1 = 30 mW) at t = 0 s, a cell chain was organized at the probe tip, which was consisted of 10 *E. colis*. Then, FP 2 was adjusted to approach the downside tip of *E. coli* 3 (indicated by the yellow dot) in the cell chain (**Figure 2a**). After turning on the laser in FP 2 (P_2 = 50 mW), the downside tip of *E. coli* 3 was trapped by FP 2 (**Figure 2b**). With a shift of 1.9 μm along the −*x* direction, the downside tip was shifted along the −*x* direction, while the upside tip (indicated by the red dot) remained stationary (**Figure 2c**). After that, *E. coil* 3 was rotated clockwise and gradually orientated along the axial direction of FP 2 (**Figure 2d**). Then, it was removed from the cell chain and shifted along the −*x* direction. After that, FP 2

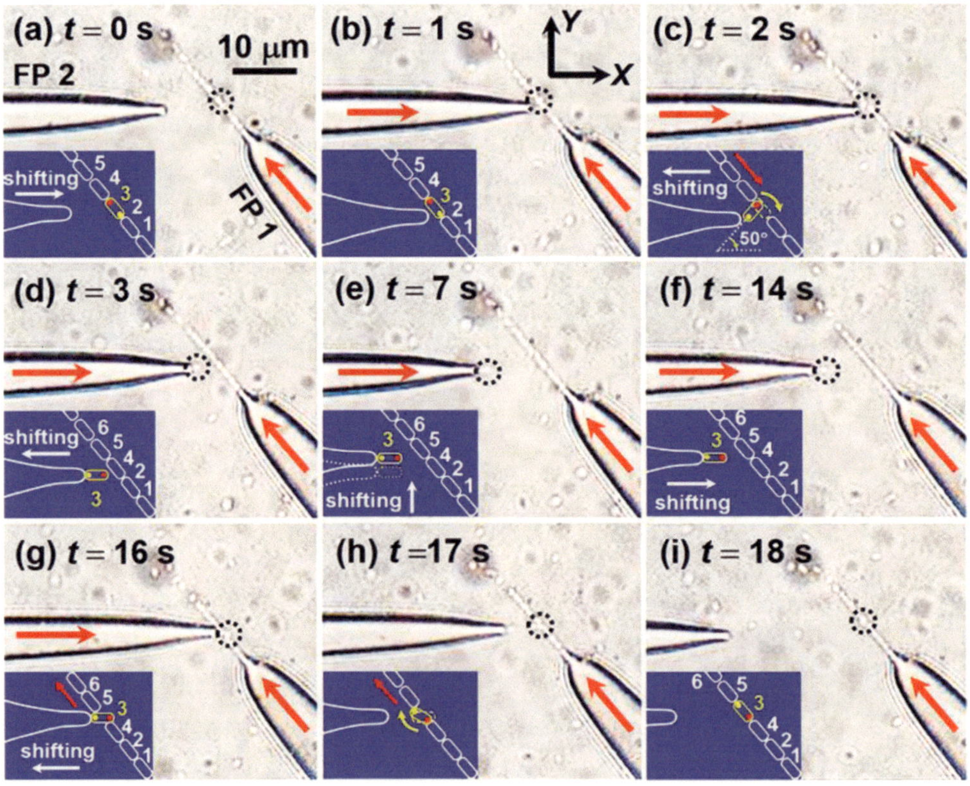

Figure 2.
Optical microscopic images of adjusting the cell contact sequence. (a) FP 2 was adjusted to approach E. coli 3. (b–d) E. coli 3 was rotated and then trapped by FP 2. (e and f) E. coli 3 was shifted with FP 2 along the +y and +x direction. (g–i) The E. coli 3 was added back into the cell chain. The two tips of E. coil 3 were indicated by yellow and red dots [38].

was shifted along the $+y$ direction with a distance of ~1 μm (**Figure 2e**), followed by a shift of 9 μm along the $+x$ direction (**Figure 2f**). Then, *E. coli* 3 was added back into the chain between *E. coils* 4 and 5 (**Figure 2g**). After the laser off in FP 2, *E. coli* 3 was reorientated along the axial direction of FP 1, and the contact sequence was changed to 1-2-4-3-5-6-7-8-9-10 (**Figure 2h** and **i**).

To quantitatively interpret the above experiment, the optical torques (T) on the *E. colis* were calculated in the regulation process, which is defined as:

$$\boldsymbol{T} = \int \boldsymbol{r}_i \times d\boldsymbol{F}_{EM} \tag{1}$$

where $\boldsymbol{r}_i$ is the position vector pointing from the central point of *E. coli* to the interaction point, $d\mathbf{F}_{EM}$ is the electromagnetic force element exerted on the interaction point i. $\boldsymbol{F}_{EM}$ can be expressed as:

$$\boldsymbol{F}_{EM} = \oint_S (\langle \boldsymbol{T}_M \rangle \cdot \boldsymbol{n}) \mathrm{d}S \tag{2}$$

where dS is the surface element surrounding the cell, $\boldsymbol{n}$ is the surface normal vector, and $\langle \boldsymbol{T}_M \rangle$ is the time-independent Maxwell stress tensor. The optical torque along the $+z$ direction is defined as positive, under which the cell will be rotated counterclockwise. As shown in **Figure 3**, the torques are positive and negative in the regions I ($-50° < \theta < 0°$) and II ($0° < \theta < 50°$). Thus, *E. coli* 3 will be rotated counterclockwise and clockwise, as indicated by the red and blue arrows, respectively. Finally, *E. coli* 3 will be oriented along the axis direction of FP 2, that is, at

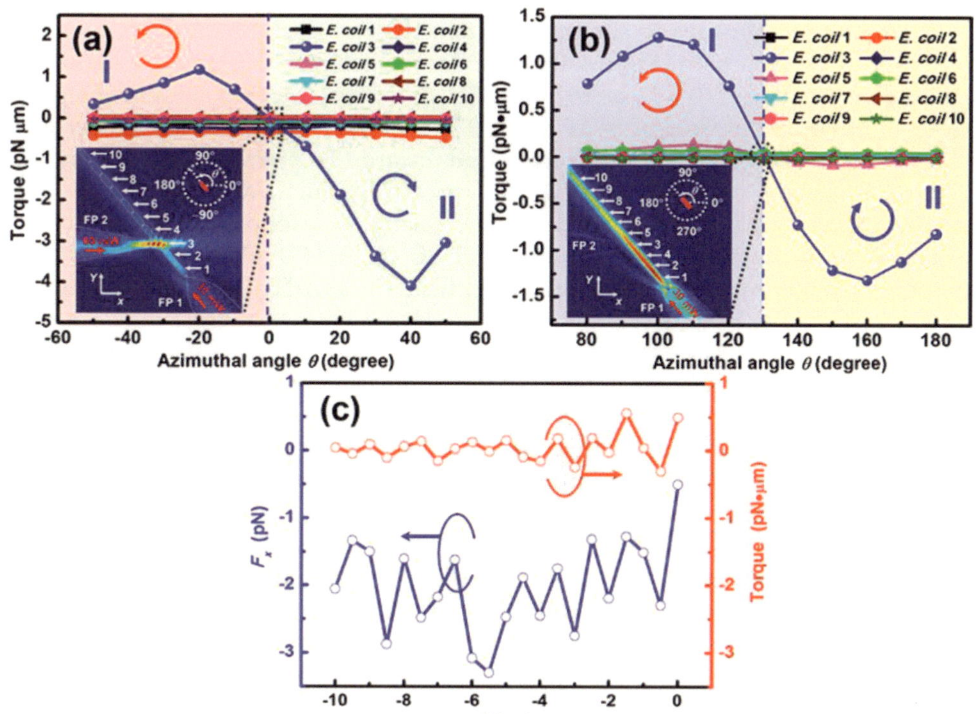

Figure 3.
Calculated optical torque and force during the regulation progress. (a) Calculated optical torque exerted on the 10 E. colis as a function of θ (FP 1: 30 mW and FP 2: 50 mW). The inset shows the energy density distribution at θ = 0°. (b) Calculated optical torque exerted on the 10 E. colis as a function of θ (FP 1: 30 mW and FP 2: 0 mW). The inset shows the energy density distribution at θ = 130°. (c) Calculated optical force F_x and torque on the E. coli 3 as a function of X [38].

$\theta = 0°$. **Figure 3b** shows the torques exerted on *E. colis* after turning off the laser in FP 2. The torques were positive and negative in the region I ($80° < \theta < 130°$) and II ($130° < \theta < 180°$), respectively. At $\theta = 130°$, the torque exerted on *E. coli* 3 was zero so that *E. coli* 3 will be stably trapped and oriented along the axial direction of FP 1. Moreover, during the shift process of FP 2, the optical force and torque exerted on *E. coli* 3 were calculated as shown in **Figure 3c**. During the shift process, the optical force was negative while the torque remained to be zero, indicating *E. coli* 3 will be stably trapped and shifted with FP 2.

Further, the numerical simulations show that the method can be used for the regulation of cell chains consisted of cells with different sizes and shapes (e.g., spherical). After the FPs incorporated into lap-on-chip platforms, the presented regulation method is expected to enable a new opportunity for the investigation of cell growth, intercellular singling pathway, and pathogenic processes.

3.2 Optofluidic organization and transport of the cell chain

Except for the precise regulation, the organized cell chain can also be dynamically transported with an optofluidic strategy, by implanting a large-tapered-angle fiber probe (LTAP) into the microfluidic technique. As shown in **Figure 4a**, when an *E. coli* cell was delivered toward the LTAP tip by the flow, it will suffer from the opposite optical scattering force. The cell kept slowing down until it was trapped stably. Then, *E. colis* were trapped one after another, and a cell chain was organized along the optical axis of LTAP (i.e., x-axis). Further, the distance between the probe tip and cell chain (D) can be dynamically adjusted so that the cell chain can be under bidirectional transportation.

Then, the experiment was conducted to demonstrate the bidirectional transportation of the cell chain (**Figure 4b**). The flow velocity was fixed at $V = 10$ μm/s, while the laser power (P) was set to be 75 mW. At $t = 0$ s, a chain consisted of eight cells was organized in front of LTAP at $D = 27$ μm. After P decreased from 75 to 60, 45, 30, and 15 mW, the chain was transported toward LTAP with D varied from 27 to 22, 17, 10, and 0 μm (**Figure 4b2–6**), respectively. Further, the cell chain can be pushed away from the probe tip by increasing the laser power. As shown in **Figure 4c**, with P increased from 15 to 30, 45, 60, 75, and 90 mW (**Figure 4b2–6**), the cell chain was continuously pushed away from the LTAP tip with D varied from 0 to 10.3, 16.5, 21.4, 26.6 and 32.9 μm, respectively.

In addition, spherical eukaryotes (e.g., yeast cells) can also be organized and transported with the proposed optofluidic technique (**Figure 5a**). The flow velocity and laser power were set to be $V = 20$ μm/s and $P = 40$ mW, respectively. First, four yeast cells were trapped and then arranged into a cell chain at $t = 0$ s. Then the cell chain was transported toward the LTAP tip with V increased to be 32 μm/s. Meanwhile, D was decreased from 23 to 18 μm (**Figure 5a3**). By varying V to 17 and 13 μm/s, the cell chain was further transported with D changed to 25 and 27.5 μm, respectively. Further, the optofluidic transportation of human red blood cells (RBCs) was also conducted to investigate the potential application in the blood diagnose. The flow velocity was fixed at 10 μm/s along the $-x$ direction. Notably, an LTAP tip with a more flat facet was fabricated to ensure the manipulation stability for the larger RBCs (**Figure 5b**). After the laser beam was injected into LTAP ($P = 40$ mW), two RBCs were transported along the $-x$ direction by the flow and then trapped by the LTAP (**Figure 5b1**). At $t = 4$, 8, and 12 s, the cell number of the RBC chain was increased from 2 to 3, 4, and 5, respectively. After organizing cell chain, it can be transported by adjusting the laser power. As shown in **Figure 5c**, a cell chain was organized in front of the probe tip. With P increased from 30 to 55 mW, D was also varied from 8.2 to 21.2 and 35.2 μm, as shown in **Figure 5c2** and **c3**, respectively. After that, the chain

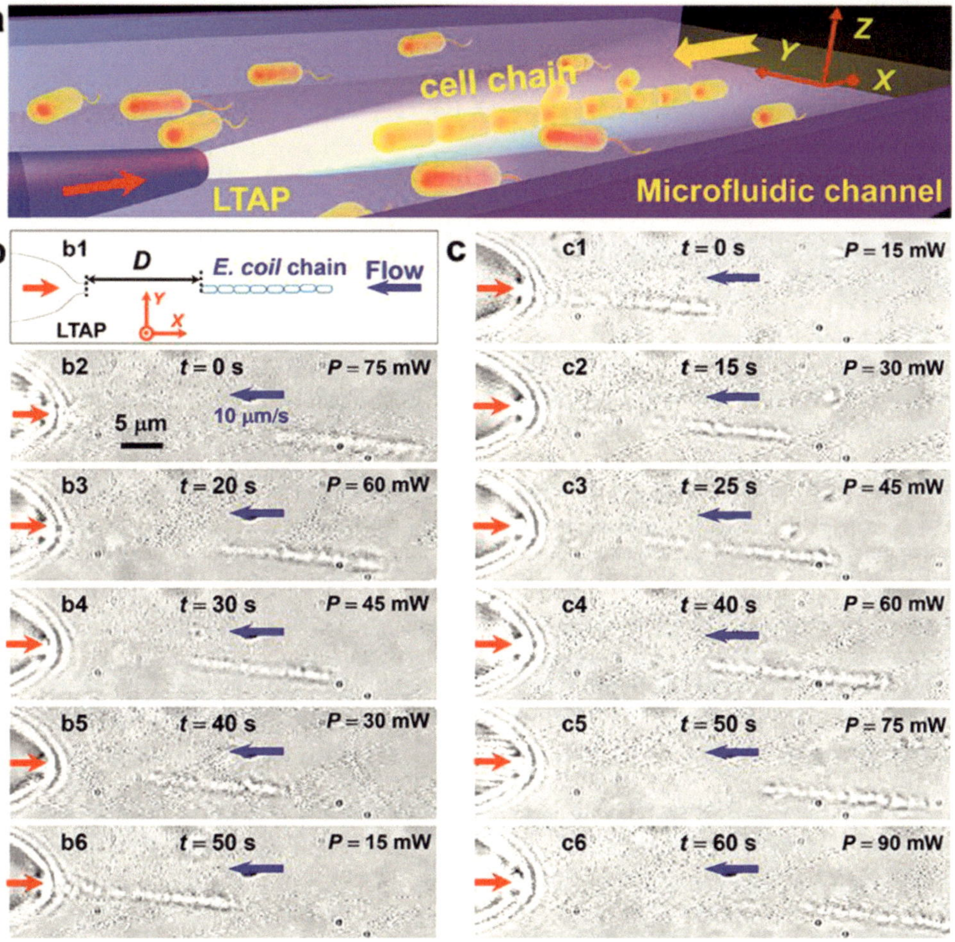

Figure 4.
(a) Schematic of optofluidic organization and transport of cell chain. (b) The cell chain was transported toward the LTAP tip by decreasing the laser power. (c) The cell chain was pushed away from LTAP tip by increasing the laser power [39].

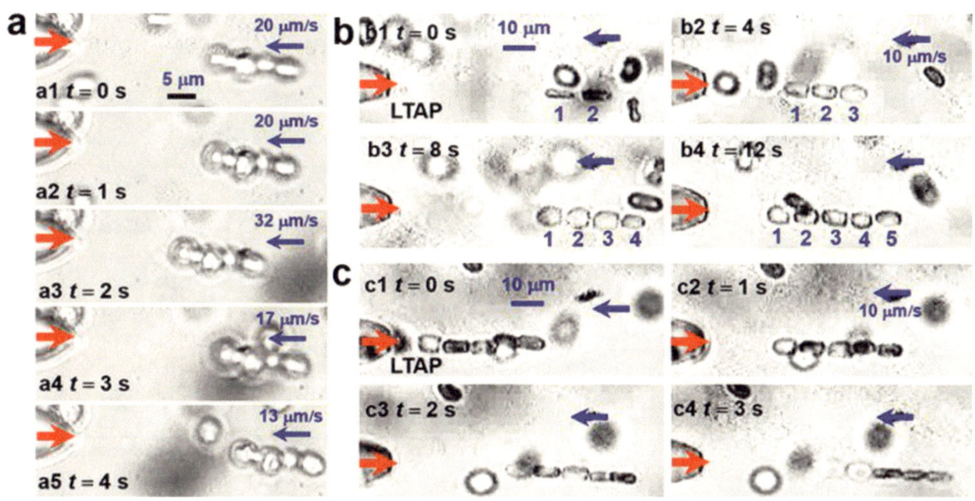

Figure 5.
(a) Optofluidic transport of yeast cell chain under various flow velocities. (b) Optofluidic organization of RBC chain consisted of five red blood cells. (c) Optofluidic transport of the RBC chain [39].

kept stationary with D remained at 35.2 μm (**Figure 5c4**). Besides, the distance can also be changed by adjusting the flow velocity, which was similar to the case of yeast cell chain. These results demonstrated that various cell chains can be dynamically transported by adjusting the laser power or flow velocity.

3.3 Optical rotation and deformation of human red blood cells

Except for the bidirectional transportation of the RBC chain, FPs were also investigated to conduct the multifunctional rotation and deformation of human red blood cells, which were of great physiological and pathological significance. As shown in **Figure 6**, one RBC was bound to the tip of TFP 1 at $\theta = 0°$ due to the optical gradient force and the Van der Waals force. At $t = 0$ s, the laser beam at a wavelength of 980 nm was injected into TFP 2 at $P_2 = 24$ mW (**Figure 6a1**). Meanwhile, the upper part of the RBC was trapped by TFP 2. By shifting TFP 2 along the $-y$ direction, the cell rotated around the x-axis with θ increased (**Figure 6a2–5**). At $t = 10$ s, TFP 2 was shifted along the $+y$ direction, and θ decreased to be 0° (**Figure 6a6–10**).

To quantitatively analyze the above rotation process, y_{TFP2} and θ were achieved as shown in **Figure 6b**. In the region I (from $t = 0$ to 3.8 s), TFP 2 was shifted along the $-y$ direction with a distance of 3.3 μm, and θ increased from 0 to 90°. Then, θ remained to be 90° from 3.8 to 5 s (region II), which indicated a certain orientation of cell can be realized by manipulating TFP 2. After that, θ increased from 90 to 172° (region III) with TFP 2 shifted along the $-y$ direction again, and then it remained to be 172° from $t = 7$ to 9.5 s (region VI). From $t = 9.5$ to 15 s, TFP 2 was shifted along the $-y$ direction so that θ decreased from 172 to 0° (region V). Thus, a controllable rotation and orientation of the RBC around x-axis can be realized by manipulating TFP 2. In addition, the shift velocity of TFP 2 (V_{TFP2}) and angular velocity of the RBC (ω_{RBC}) were also calculated. As shown in **Figure 6c**, the value of ω_{RBC} monotonously varied with V_{TFP2}, with a maximum of 4.7 rad/s. Notably, the trapped part can be changed by adjusting TFP 2 along the z direction. For example,

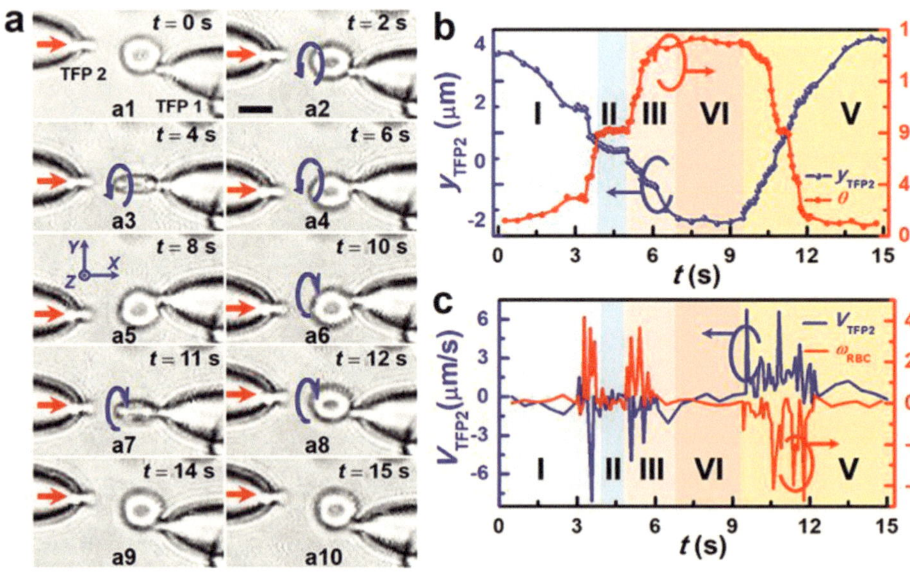

Figure 6.
RBC rotation around x-axis. (a) Optical microscopic images for rotating an RBC around x-axis. Scale bar: 5 μm. (b) Calculated y_{TFP2} and θ in the rotation process. (c) Calculated V_{TFP2} and ω_{RBC} in the rotation process [40].

by trapping the left part of the cell, the cell will rotate around z-axis, which can realize the multifunctional rotation around different axes.

Further, a stretch of single or multiple RBCs can also be realized by using two TFPs. As schematically shown in **Figure 7a**, after the laser beams injected into both TFPs 1 and 2, three RBCs are trapped and then stretched along the optical axis of TFPs. The experiments were then conducted to demonstrate the above stretch mechanism. At t = 0 s, two RBCs were located between two TFPs, with diameters of 5.6 and 6.4 μm, respectively (**Figure 7b1**). Then, two laser beams at the wavelength of 980 nm were injected into the TFPs with the power of $P_1 = P_2 = 20$ mW. After that, the RBCs moved toward each other and finally became contacted due to the optical force (**Figure 7b2**). Meanwhile, they were gradually stretched till reaching an equilibrium at t = 10 s (**Figure 7b3**). The deformation degree of RBCs can be described by the shear strain (γ) which was defined as $\gamma = \Delta l/l$, where Δl and l are the cell stretch length and original length, respectively. At t = 10 s (**Figure 7b3**), the shear strain of two cells reached the maximum of 0.14 and 0.12, respectively. Then, they kept stretched until the laser was turned off at t = 35 s (**Figure 7b4**). After that, the RBCs were gradually resumed their original shapes (**Figure 7b5, 6**).

Similarly, the simultaneous stretch of three RBCs was also conducted, as shown in **Figure 7c**. The diameters of RBCs 1, 2, and 3 were 6.7, 5.7, and 6.9 μm, respectively. At t = 2 s, the RBCs began to be trapped and then stretched after the laser turned on (**Figure 7c2**). Meanwhile, the distance between the cells was decreased, while γ were increased. They gradually became contacted with each other. At t = 5 s,

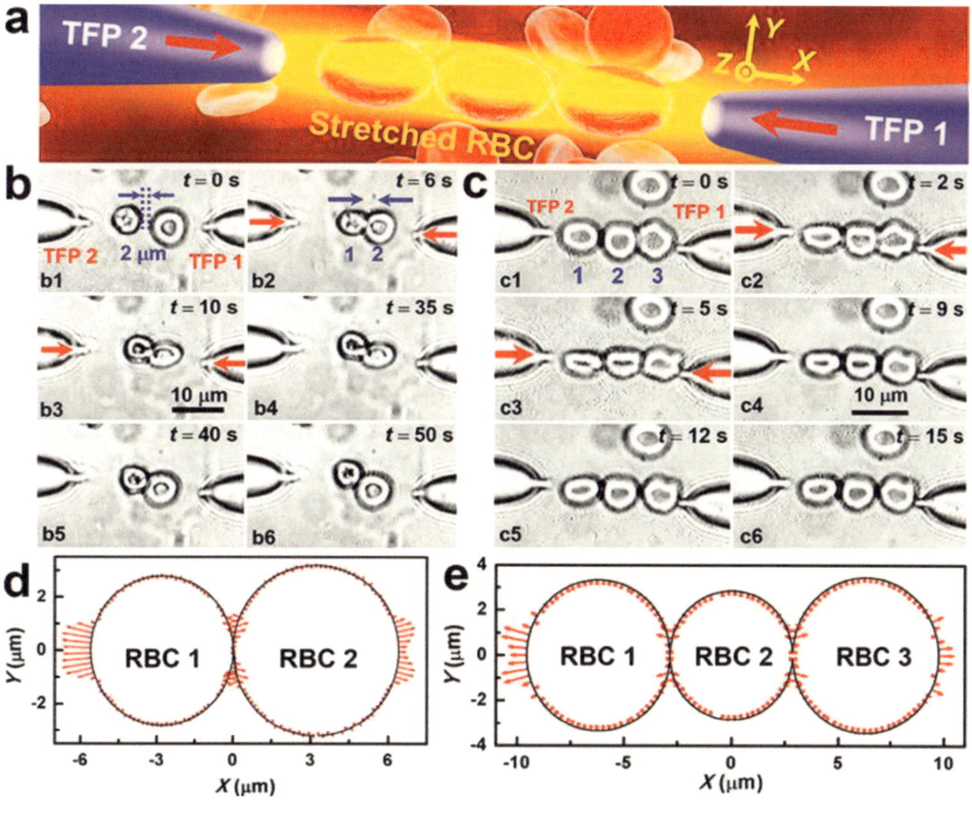

Figure 7.
The stretch of multiple RBCs with light from two TFPs. (a) Schematic of stretching multiple RBCs with two TFPs. (b) Optical microscopic images of stretching two RBCs. (c) Optical microscopic images of stretching three RBCs. (d) Stress distribution on the surfaces of two RBCs. (e) Stress distribution on the surfaces of three RBCs [40].

the three cells reached the equilibrium with $\gamma = 0.15$, 0.1, and 0.12 (**Figure 7c3**). After that, RBCs started to resume their original shapes by turning off the laser at $t = 9$ s (**Figure 7c4–6**). Further, the normalized stress distribution on the RBC surfaces was also investigated, as indicated by the red arrows in **Figure 7e** and **f**. The stress was mainly distributed along the x-axis, and the stress direction was toward the outside of cells. Therefore, the RBCs were stretched and further became contacted with each other.

4. Discussion

From the above analysis, it can be seen that the multifunctional trap and manipulation can be conducted with the fiber probe-based optical tweezers. As we know, a precise control of cell behavior was essential in the research of the endocytosis and exocytosis process, especially for nanomedicine injection, intracellular signaling pathway, and pathogenic progress. After cells manipulated in a controlled manner, various microparticles or viruses can be brought into contact with the targeted cells at specific well-defined time points and positions. Meanwhile, the spatiotemporal effect can be quantitatively investigated of different extracellular cues on the endocytosis and exocytosis process.

Then the potential application and advantages were discussed for fiber probe-based tweezers on the endocytosis and exocytosis purpose. As for the nanomedicine injection, the cells can be located at a different distance to the medicine with the assistance of fiber probes, providing a great way to study the diffusion and transportation process. Further, the nanomedicine can be adjusted to approach various sites of the cell membrane. Then, the endocytosis efficiency on the interaction sites can be quantitatively investigated. Meanwhile, various nanomedicines can approach the targeted cell simultaneously with controlled sites, providing an insight into the study of selective phagocytosis progress. While for the intracellular signaling pathway, spatial manipulation will be beneficial to analyze the effect of contact distance on the cell interaction. As we know, the cell can interact with each other through the exchange of soluble signaling molecules or direct cell-cell contact, which can vary in both time and space continuously. Thus, it is of great importance in dynamically adjusting the cell-cell interaction distance and contact sequence, which can be conducted by using the fiber-probe tweezers. Further, the detailed pathway for the intracellular signaling can be individually investigated with the proposed technique. Besides, in the pathogenic progress, the specific infection site can be decided by using fiber probes to manipulate pathogenic bacterium and targeted cell simultaneously. Thus, the dependence of the contact site on the infection performance can be investigated, which opens up the possibility for analyzing pathogenic dynamics not accessible through passive observation.

5. Conclusions

In this chapter, the fiber probe has been demonstrated to conduct the multifunctional manipulation of cell chains, including the sequential organization, precise regulation, and bidirectional transportation. Besides, the dynamic rotation and deformation of the human red blood cell were also realized using the fiber probes. With the advantage of ease integration, high flexibility, and noninvasive, the proposed technique can provide a great perspective in the investigation of the endocytosis and exocytosis purpose. The targeted cells and pathogenic bacterium can be dynamically manipulated simultaneously, which is anticipated to be useful in the analysis, diagnosis, and treatment of cell diseases. By incorporating the FPs into lap-on-chip platforms, the presented technique is expected to enable a new opportunity for the investigation of the nanomedicine injection, intracellular signaling pathway, and pathogenic process.

Acknowledgements

This work was supported by the National Natural Science Foundation of China (No. 11774135, 11874183, and 61827822), the Tip-top Scientific and Technical Innovative Youth Talents of Guangdong Special Support Program (No. 2015TQ01X267), the Fundamental Research Funds for the Central Universities (No. 21618301), and the PhD Start-up Fund of Natural Science Foundation of Guangdong Province (No. 2018A030310501).

Conflict of interest

The authors declare no competing financial interests.

Author details

Xiaoshuai Liu and Yao Zhang*
Institute of Nanophotonics, Jinan University, Guangzhou, China

*Address all correspondence to: zhyao5@jnu.edu.cn

References

[1] Oh N, Park JH. Endocytosis and exocytosis of nanoparticles in mammalian cells. International Journal of Nanomedicine. 2014;**9**:51-63. DOI: 10.2147/IJN.S26592

[2] Aoki SK, Pamma R, Hernday AD, Bickham JE, Braaten BA, Low DA. Contact-dependent inhibition of growth in *Escherichia coli*. Science. 2005;**309**:1245-1248. DOI: 10.1126/science.1115109

[3] Aoki SK, Diner EJ, Roodenbeke CK, Burgess BR, Poole SJ, Braaten BA, et al. A widespread family of polymorphic contactdependent toxin delivery systems in bacteria. Nature. 2010;**468**:439-442. DOI: 10.1038/nature09490

[4] Tang J, Peng R, Ding J. The regulation of stem cell differentiation by cell-cell contact on micropatterned material surfaces. Biomaterials. 2010;**31**:2470-2476. DOI: 10.1016/j.biomaterials.2009.12.006

[5] Charest JL, Jennings JM, King WP, Kowalczyk AP, Garcia AJ. Cadherin-mediated cell-cell contact regulates keratinocyte differentiation. The Journal of Investigative Dermatology. 2009;**129**:564-572. DOI: 10.1038/jid.2008.265

[6] Krauss RS, Cole F, Gaio U, Takaesu G, Zhang W, Kang JS. Close encounters: Regulation of vertebrate skeletal myogenesis by cell-cell contact. Journal of Cell Science. 2005;**118**:2355-2362. DOI: 10.1242/jcs.02397

[7] Castillo J, Dimaki M, Svendsen WE. Manipulation of biological samples using micro and nano techniques. Integrative Biology. 2009;**1**:30-42. DOI: 10.1039/b814549k

[8] Wlodkowic D, Cooper JM. Microfabricated analytical systems for integrated cancer cytomics. Analytical and Bioanalytical Chemistry. 2010;**398**:193-209. DOI: 10.1007/s00216-010-3722-8

[9] Saltzman WM, Olbricht WL. Building drug delivery into tissue engineering. Nature Reviews. Drug Discovery. 2002;**1**:177-186. DOI: 10.1038/nrd744

[10] Chen CS, Tan J, Tien J. Mechanotransduction at cell-matrix and cell-cell contacts. Annual Review of Biomedical Engineering. 2004;**6**:275-302. DOI: 10.1146/annurev.bioeng.6.040803.140040

[11] Hidalgo-Carcedo C, Hooper S, Chaudhry SI, Williamson P, Harrington K, Leitinger B, et al. Collective cell migration requires suppression of actomyosin at cell-cell contacts mediated by DDR1 and the cell polarity regulators Par3 and Par6. Nature Cell Biology. 2011;**13**:49-58. DOI: 10.1038/ncb2133

[12] Kirschbaum M, Guernth-Marschner CR, Cherré S, Peña AP, Jaeger MS, Kroczek RA, et al. Highly controlled electrofusion of individually selected cells in dielectrophoretic field cages. Lab on a Chip. 2012;**12**:443-450. DOI: 10.1039/c1lc20818g

[13] Hamdi FS, Français O, Subra F, Dufour-Gergam E, Pioufle BL. Microarray of non-connected gold pads used as high density electric traps for parallelized pairing and fusion of cells. Biomicrofluidics. 2013;**7**:044101. DOI: 10.1063/1.4813062

[14] Kimura T, Sato Y, Kimura F, Iwasaka M, Ueno S. Micropatterning of cells using modulated magnetic fields. Langmuir. 2005;**21**:830-832. DOI: 10.1021/la047517z

[15] Sakar MS, Steager EB, Kim DH, Kim MJ, Pappas GJ, Kumar V. Single

cell manipulation using ferromagnetic composite microtransporters. Applied Physics Letters. 2010;**96**:043705. DOI: 10.1063/1.3293457

[16] Khademhosseini A, Yeh J, Jon S, Eng G, Suh KY, Burdick JA, et al. Molded polyethylene glycol microstructures for capturing cells within microfluidic channels. Lab on a Chip. 2004;**4**:425-430. DOI: 10.1039/b404842c

[17] Théry M. Micropatterning as a tool to decipher cell morphogenesis and functions. Journal of Cell Science. 2010;**123**:4201-4213. DOI: 10.1242/jcs.075150

[18] Huang NT, Zhang HL, Chung MT, Seo JH, Kurabayashi K. Recent advancements in optofluidics-based single-cell analysis: optical on-chip cellular manipulation, treatment, and property detection. Lab on a Chip. 2014;**14**:1230-1245. DOI: 10.1039/c3lc51211h

[19] Yi CQ, Li CW, Ji SL, Yang MS. Microfluidics technology for manipulation and analysis of biological cells. Analytica Chimica Acta. 2006;**560**:1-23. DOI: 10.1016/j.aca.2005.12.037

[20] Schmidt BS, Yang AH, Erickson D, Lipson M. Optofluidic trapping and transport on solid core waveguides within a microfluidic device. Optics Express. 2007;**15**:14322-14334

[21] Psaltis D, Quake SR, Yang CH. Developing optofluidic technology through the fusion of microfluidics and optics. Nature. 2006;**442**:381-386

[22] Dholakia K, Reece P, Gu M. Optical micromanipulation. Chemical Society Reviews. 2008;**37**:42-55. DOI: 10.1039/b512471a

[23] Daly M, Sergides M, Chormaic SN. Optical trapping and manipulation of micrometer and submicrometer particles. Laser & Photonics Reviews. 2015;**9**:309-329. DOI: 10.1002/lpor.201500006

[24] Phillips DB, Padgett MJ, Hanna S, Ho YD, Carberry DM, Miles MJ, et al. Shape-induced force fields in optical trapping. Nature Photonics. 2014;**8**:400-405. DOI: 10.1038/NPHOTON.2014.74

[25] Rodríguez-Sevilla P, Rodríguez-Rodríguez H, Pedroni M, Speghini A, Bettinelli M, Solé JG, et al. Assessing single upconverting nanoparticle luminescence by optical tweezers. Nano Letters. 2015;**15**:5068-5074. DOI: 10.1021/acs.nanolett.5b01184

[26] Kim K, Yoon J, Park YK. Simultaneous 3D visualization and position tracking of optically trapped particles using optical diffraction tomography. Optica. 2015;**2**:2334-2536. DOI: 10.1364/OPTICA.2.000343

[27] Grier DG. A revolution in optical manipulation. Nature Photonics. 2003;**424**:21-27. DOI: 10.1038/nature01935

[28] Xin HB, Xu R, Li BJ. Optical trapping driving, and arrangement of particles using a tapered fibre probe. Scientific Reports. 2012;**2**:1-8. DOI: 10.1038/srep00818

[29] Padgett M, Leonardo RD. Holographic optical tweezers and their relevance to lab on chip devices. Lab on a Chip. 2011;**11**:1196-1205. DOI: 10.1039/c0lc00526f

[30] Kirkham GR, Britchford E, Upton T, Ware J, Gibson GM, Devaud Y, et al. Precision assembly of complex cellular microenvironments using holographic optical tweezers. Scientific Reports. 2015;**5**:8577. DOI: 10.1038/srep08577

[31] Akselrod GM, Timp W, Mirsaidov U, Zhao Q, Li C, Timp R, et al. Laser-guided assembly of heterotypic three-dimensional living cell microarrays.

Biophysical Journal. 2006;**91**:3465-3473. DOI: 10.1529/biophysj.106.084079

[32] Righini M, Ghenuche P, Cherukulappurath S, Myroshnychenko V, Abajo FJ, Quidan R. Nano-optical trapping of Rayleigh particles and *Escherichia coli* bacteria with resonant optical antennas. Nano Letters. 2009;**9**:3387-3391. DOI: 10.1021/nl803677x

[33] Juan ML, Righini M, Quidant R. Plasmon nano-optical tweezers. Nature Photonics. 2011;**5**:349-356. DOI: 10.1038/nphoton.2011.56

[34] Decombe JB, Valdivia-Valero FJ, Dantelle G, Leménager G, Gacoin T, Francs GC, et al. Luminescent nanoparticle trapping with far-field optical fiber-tip tweezers. Nanoscale. 2016;**8**:5334-5342. DOI: 10.1039/c5nr07727c

[35] Li YC, Xin HB, Liu XS, Li BJ. Non-contact intracellular binding of chloroplasts in vivo. Scientific Reports. 2015;**5**:10925. DOI: 10.1038/srep10925

[36] Fazal FM, SM B. Optical tweezers study life under tension. Nature Photonics. 2011;**5**:318-321. DOI: 10.1038/nphoton.2011.100

[37] Gross SP. Application of optical traps in vivo. Methods in Enzymology. 2003;**361**:162-174. DOI: 10.1016/S0076-6879(03)61010-4

[38] Liu XS, Huang JB, Zhang Y, Li BJ. Optical regulation of cell chain. Scientific Reports. 2015;**5**:11578. DOI: 10.1038/srep11578

[39] Liu XS, Huang JB, Li YC, Zhang Y, Li BJ. Optofluidic organization and transport of cell chain. Journal of Biophotonics. 2017;**10**:1627-1635. DOI: 10.1002/jbio.201600306

[40] Liu XS, Huang JB, Li YC, Zhang Y, Li BJ. Rotation and deformation of human red blood cells with light from tapered fiber probes. Nano. 2017;**6**:309-316. DOI: 10.1515/nanoph-2016-0115

Section 3

Modern Aspects on Optical Fibers

Chapter 3

Structured Light Fields in Optical Fibers

Monika Bahl

Abstract

Structured light, tailored light, sculpted light, or shaped light is a term used for custom light fields and finds enormous use in literature these days. Some of the history's most brilliant researchers, from Newton to Maxwell to Einstein, have studied the nature of light over the centuries. We believe that we know everything about light, its generation, detection, and applications; yet, it continues to surprise us even today. Indeed, one discovery about light's peculiar behavior has offered a new insight into how light works and rendered some intriguing applications. In 1992, physicists mastered a surprising feat—generating light beams that twist like a helical corkscrew. This phenomenon is called twisted light and has led to an altogether new field of optics, known as singular optics. Today, twisted light is being used to build optical tweezers and ultra-powerful microscopes, and it could eventually be used in microscale machinery and for novel spectroscopic analyses. But perhaps the most important use of this structured light is in optical communications, where it moves through optical fibers. This light has the potential to greatly enhance the bandwidth of data networks and, hence, the speed of data transmission.

Keywords: structured beams, tailored light, optical vortex, orbital angular momentum, optical fibers, data communication

1. Introduction

Light is an electromagnetic wave, composed of electric and magnetic fields. The fields oscillate in a direction perpendicular to the direction in which the wave is moving. If the electric field is always oscillating in the same plane, the light is said to be linearly polarized. Photons of such light possess a linear momentum. Such photons have the power to propel a boat if the solar sails absorb the linear momentum. If the direction of the plane in which the light's electric field is vibrating is itself rotating as the wave moves, light is said to be circularly polarized. In this case, light is said to possess a spin angular momentum. When such a light hits a floating ball, it will start to spin like a planet, rotating about its own axis. We thus had these two types, and surprisingly, it turned out that these are not the only ways light behaves. Allen et al., in 1992, showed in his seminal paper that light may possess another strikingly different characteristic behavior [1]. It happens when the wave fronts, instead of moving in a straight line or diverging/converging, tend to bend, rotate, and propagate in a helical fashion. This was something unexpected and very different from the already known phenomenon. This implied that the energy propagation direction is not a straight line but that too forms a helical trace as the wave moves in the forward direction. Hence, the Poynting vector ($S = E \times H$) changes its direction continuously as the wave moves. Therefore, a given phase front will rotate around the center and trace a helix as it propagates. Such a wave is said to possess a momentum, apart from the linear and spin momentum, which is termed as the orbital angular momentum or OAM. The center of the wave thus carries a singularity of phase called the optical vortex, and the intensity profile represents a doughnut [2–17]. Hitting with this type of light, a free-floating ball will revolve in a circle about a central point as a planet orbits a star. Thus, a "vortex beam" represents a column of light with a hole in the center. In 1991, physicist Robert Spreeuw, at the time a PhD student in Han Woerdman's lab at Leiden University in The Netherlands, sat down during a team coffee break and presented some ideas about how to make twisted light. "The first reactions were a bit doubtful," Spreeuw recollects. "But we kept thinking about it and, bit by bit, it started to look more realistic."

Angular momentum of photons consists of two different components. The first one is the spin angular momentum (SAM), which corresponds to the polarization of the photon. The second component is the orbital angular momentum (OAM), which relates to the spatial phase profile of the photon." Both the components have been used extensively in optical experiments in the laboratories. Moreover, polarization has been used successfully in quantum experiments in free space for about an order of 100 km [18–23]. The polarization of a photon is still more easily controllable and resistant against atmospherical influences and resides in a two-dimensional state space. This places an inherent limit on how much information one can send per photon. An alternate way to encode information is in the OAM, another degree-of-freedom of a photon that offers infinite unbounded number of discrete levels theoretically and is able to render faster effective communication over long distances [18–23].

A brief note on the light-carrying OAM would give an insight into what it looks like. This light has a "twisted" or helical wave front, with an azimuthal phase varying from 0 to $2\pi l$. The integer l stands for the topological charge or helicity, and $l\hbar$ is the OAM of the photon. In 1992, Woerdman and his colleague Les Allen created twisted light in the lab and showed that even a single photon of light has OAM [1]. In the next year, they showed how to convert a normal helium-neon laser to one that carried OAM. One way to create twisted light is to send it through a phase plate with thickness varying azimuthally. This shapes the wavefonts of the field into a form resembling a helix. It is as if you took a rod and swirled it around to create a vortex in the phases of the electromagnetic waves. This is similar to a vortex in water waves. Topological charge (l) 1 means there is 1 helix propagating clockwise (say) in space, charge 2 means 2 interwined helices propagate, while −1, −2, etc., mean that 1 and 2 helices propagate anticlockwise in space. Higher order charges would increase the number of helices. **Figure 1** shows a plane wave ($l = 0$) and helical waves with $l = 1$ and −1, respectively.

Beams carrying OAM can also be generated in the lab using other techniques like a helical mirror or using a computer-generated hologram. Although many of these and other techniques are available, researchers are excited about the potential uses of these beams. Such beams have the potential to trap and rotate particles (optical tweezers), render fast optical communication, used for quantum computing, and use in various other areas. In 1995, an Australian team placed small particles in the dark, central cavity of an OAM laser and watched them whirl around, providing visual proof that the light was carrying OAM. The researchers could even reverse the direction of the OAM laser's twist and spin the particles the opposite way. Thus, an exciting new degree of freedom has been made available to the researchers to extract the potential of these twisted wavefronts. These can be launched into the optical fibers (OF) and photonic crystal fibers (PCF) and provide an all new technique to communicate data at speeds higher than the available ones. But before

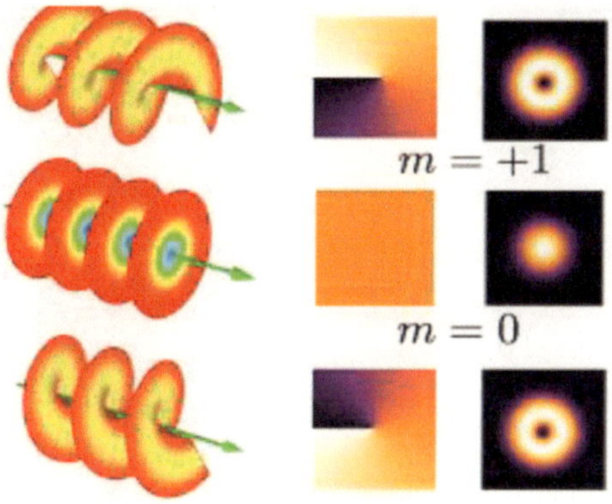

Figure 1.
Phase fronts/wave fronts, phase profiles and transverse intensity profiles of a positive helical wave with m = 1, a plane wave, and a negative helical wave with m = −1. (ref: phot https://futurism.com/new-understanding-twisted-light-affect-use-quantum-computing).

moving further, let me give you a brief description of what optical fibers are and how they are used in data communication.

2. What are optical fibers?

Optical fibers are dielectric wave-guides, circular in shape that can transport optical information and energy. They have a central core that is surrounded by a concentric cladding that has a slightly lower refractive index than that of the core. Optical fibers are normally made of silica doped with index-modifying dopants such as GeO_2 [24–29]. A protective coating of one or two layers of a cushioning material (such as acrylate) is used to reduce cross talk between adjacent fibers and microbending, which occurs when fibers are pressed against rough surfaces. Microbending and cross talk increase the loss of optical energy as the optical beams propagate through the fiber. Fibers are typically incorporated into cables in order to provide environmental protection. Typical cables have a polyethylene sheath that encases the fiber within a strength member such as steel or Kevlar strands. **Figure 2** shows a typical sketch of an optical fiber.

Since the core has a higher index of refraction than the cladding, light will be confined to the core when the condition for total internal reflectance (TIR) is

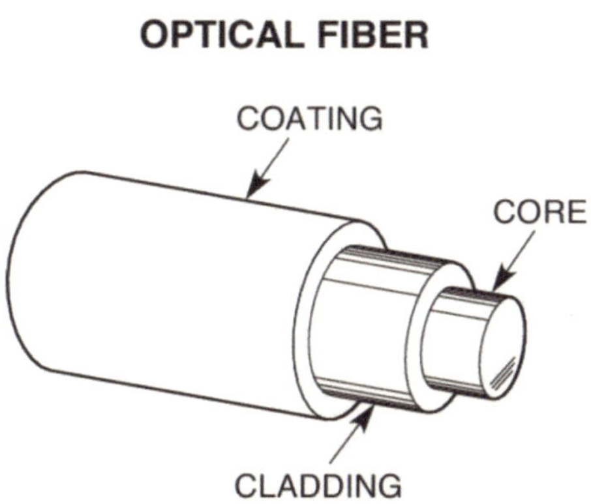

Figure 2.
Cross-section view of an optical fiber (https://www.newport.com/t/fiber-optic-basics).

Figure 3.
Close-up view of a fiber preform (https://www.newport.com/t/fiber-optic-basics).

Figure 4.
Fibers can be connected (https://www.newport.com/t/fiber-optic-basics).

met. The composition and geometry of the fiber determine the fiber modes or the discrete set of electromagnetic fields, which can propagate in the fiber (**Figure 3**).

Directly connectorized fibers may furthermore provide beam expansion to lower the fiber end facet intensity and reduce the risk of damage at high power levels (**Figure 4**).

3. Can optical fibers be used for communication and why?

No one desires to have a slow Internet connection! I am sure you agree with me on this. It is very frustrating to have a slow Internet speed when one wants to download something or watch a favorite movie online.

Before the wireless Internet era, there used to be the wired, dial-up Internet. In this type of connection, a phone line was used to connect to the Internet. Several hours were spent only praying for a successful connection. A fast Internet was definitely a luxury during those days.

At that point, broadband connections came in, which totally changed the Internet scene. The rates of 2–10 mbps became normal. Nonetheless, this still utilized great old copper wires as the transmission medium.

Around 10 years back, Internet suppliers started utilizing fiber as a medium to transmit signals. The purpose behind this was that the transmission happened along these lines with minimal losses, giving a lot quicker Internet speeds. In 2014, an exploration aggregate at the Technical University of Denmark (DTU) succeeded to transfer 43 terabits for each second over a solitary optical fiber with only one laser transmitter!

Therefore, what makes this fiber-optic technology lot more adept than copper wires? Let us, therefore, have a look at the two transmission media to get to an answer to this question.

3.1 Copper wire

Electrical pulses are sent through a copper wire in case of wired copper communication. How much of the signal will be retained by the time it reaches its destination is determined by the signal strength. The wire's electromagnetic field is constantly monitored for changes at the destination (for example, the router). The destination registers a "1" (logic high), when the field is strong, i.e., above a certain measurement, say, a, while a "0" (logic low) is registered if it dips below that particular measurement.

3.2 Fiber optic cable

A fiber-optic cable is made from fine hair-like glass fiber, which carries light impulses that are transmitted by an LED or a laser source. Data in optical fiber are transmitted in the form of light (**Figure** 5).

Consider a long and flexible pipe, with its insides perfectly coated with silver halide, such that the inside is all a mirror. It is like a cylindrical mirror from inside. When you flash a source of light (a laser or may be a torch), what do you see at the other end? The light will reach the other end, regardless of whether the pipe is straight, curved, or twisted. Is not it? Yes, it will. This is because light will reflect off the sides of the flexible tube at all angles with almost negligible losses. But mirror tubes would be too bulky to handle. Thus, optical fibers are used to serve the purpose. Optical fibers are such flexible pipes made of glass instead of mirrors. It employs the principle of total internal reflection to transmit light from one to the other end.

Glass is amazingly pure; light can make it through even if it is several miles long. The glass for an optical fiber is drawn into an extremely thin strand, with a thickness comparable to that of a human hair. The glass strand is then coated in two layers of plastic.

3.3 Total internal reflection in fibers

Light rays traveling from a denser medium to a rarer medium speed up at the boundary. This causes the rays to bend when they pass from glass to air at an angle

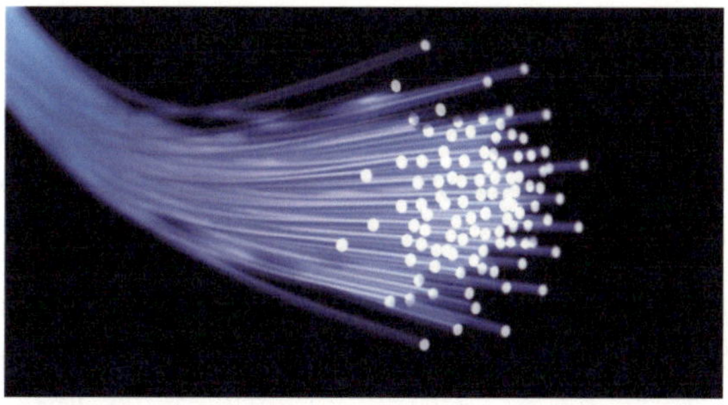

Figure 5.
Light travels in an optical fiber cable (credit: ProMotion/fotolia).

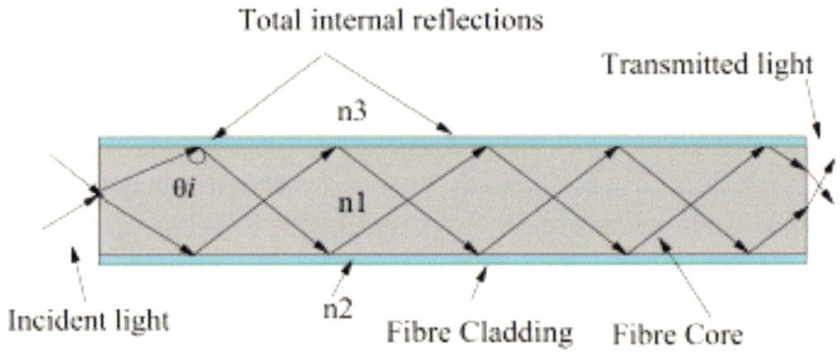

Figure 6.
Total internal reflection in an optical fiber.

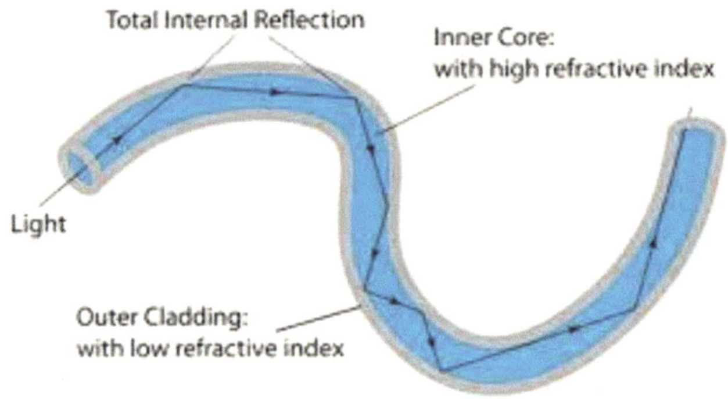

Figure 7.
Light rays undergo total internal reflection inside an optic fiber that is bent.

other than 90°. This is called refraction. The two layers of plastic form the two mediums on which the reflections occur (**Figure 6**).

Beyond a certain angle, called the critical angle, all the waves reflect back into the glass. We say that they are totally internally reflected. The light rays stay inside the optic fiber and are transmitted over long distances with negligible loss (**Figure 7**).

3.4 Benefits of fiber optics

The following properties make fiber optic cable superior to conventional copper cables.

1. Bandwidth

An optical fiber provides more bandwidth as compared to a copper wire and has a standardized performance up to 10 Gbps and beyond, something that it is impossible to achieve with a copper wire. A higher bandwidth means that the fiber can carry more information with far greater efficiency than a copper wire.

2. Range of transmission

Data travel in the form of light through a fiber optic cable. The loss of quality of signal is almost negligible in the case of total internal reflections within the glass fiber, and hence, very little signal loss occurs during transmission. Data can, thus, move at higher speeds and for greater distances.

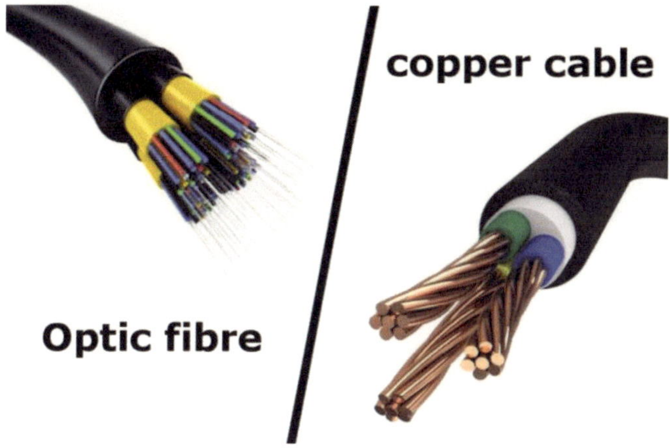

Figure 8.
Optic fiber is much lighter and thinner compared to the conventional copper cable.

3. Not susceptible to interference

Data transmitted through a fiber-optic cable is much less susceptible to noise. It is also less susceptible to electromagnetic interference as compared to one moving through a copper wire. For example, there would be almost zero degree of quality degradation in a signal through a fiber optic cable, say even for a distance of over two kilometers, while a signal transmitted through a copper wire would experience a great deal of degradation in quality. It is, therefore, so efficient that roughly 99.6% of the signal reaches the router in most cases.

4. Size, weight, and strength

A copper cable is bulky and heavy as compared to a fiber optic cable that is much lighter and thinner. It can be used very efficiently in underground pipes that are confined to the ground and are also much stronger, with eight times the pulling tension of a copper wire. It is also tough against environmental factors and atmospheric distortions and hard to damage or kink (**Figure 8**).

5. Cost

The initial cost of material and installation of an optical fiber is high as compared to a copper wire, but in the long run, the working cost is much less. Moreover, a fiber network has a low maintenance cost and requires very few networking hardware.

6. Durability

Fiber optic cable is highly resistant to many environmental factors that, otherwise, affect a copper cable network. No electric current can flow through the former as the core is made up of glass which is a perfect insulator. Fiber cables can be made to run next to any industrial equipment too. An optical fiber is also more resistant to temperature fluctuations as compared to copper and can also be submerged in water.

As discussed, optical fibers can communicate data through transmission of waves based on the phenomenon of total internal reflection. The theoretical bandwidth of optical fiber transmission is of the order of few terabits.

Bandwidth can be enhanced by employing two techniques. The first one is known as the time division multiplexing (TDM). Multiple channels are transmitted

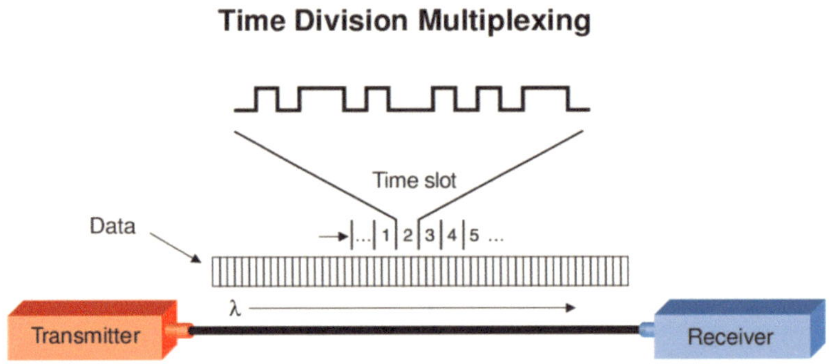

Figure 9.
Time division multiplexing (https://www.newport.com/t/fiber-optic-basics).

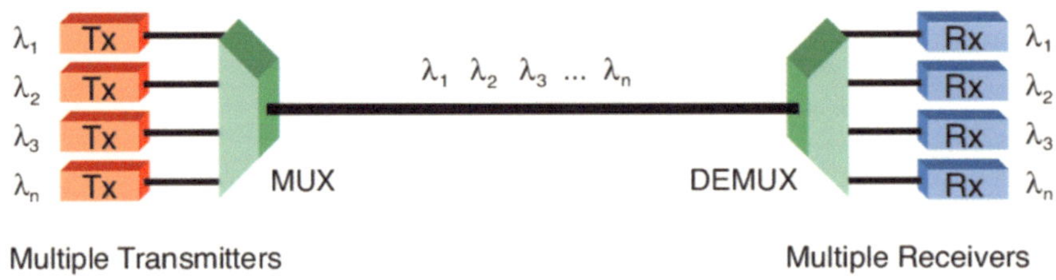

Figure 10.
Wavelength division multiplexing (https://www.newport.com/t/fiber-optic-basics).

onto a single carrier by increasing the modulation rate, and each channel is allotted a time slot (**Figure 9**).

The second method is known as wavelength division multiplexing or WDM. Using this method, capacity can be increased by using more than one optical carrier (wavelength) in a single fiber. Therefore, adding a second transmitter and receiver to an optical fiber can double the bandwidth of that communications system (**Figure 10**).

As per the data on the Newport website, the ITU (International Telecommunication Union) had proposed a set of closely spaced wavelengths in the 1550 nm window. This method of WDM is known as Dense Wavelength Division Multiplexing or DWDM. These different wavelengths or channels, spaced 100 GHz apart or 0.8 nm approximately, form the ITU-T grid. The 1550 nm window has the smallest amount of attenuation and lies in the band in which erbium-doped optical amplifiers operate. DWDM systems have a fixed starting and a distinct ending point. Thus, these are, therefore, called point-to-point links. Research is being done to make these networks evolve into completely configurable networks that are not limited to fixed point-to-point links.

Transparency in the optical layer opens many possibilities for the future. Digital and analog transmission can occur on the same fiber. Different bit rates using different protocols will all travel together. Current research is being performed on reconfiguring an optical network in real time. Wavelength selective switching allows wavelengths to be routed through the network individually. Some of the applications of this are for network restoration and redundancy, which may reduce or entirely eliminate the need for an entire back up system to help the network recover from failures such as equipment malfunctions or fiber breaks. A reconfigurable network may offer bandwidth on demand to configure itself to optimize for traffic bottlenecks. The future may also include wavelength translation to convert traffic on one wavelength to another wavelength in the optical domain.

All optical switching is still in the research phase; however, researchers are looking for ways to create reliable, low loss switches with fast switching speeds. Investigation into the possibility of optical packet switching and other novel technologies is currently underway. The all-optical network may be just around the corner.

The future is fiber-optic technology, and the network is growing exponentially worldwide. Most major companies are already using fiber-optic systems in their backbone applications. These systems offer higher reliability and speed.

4. Can structured beams propagate in optical fibers?

4.1 Yes... twisted light through the light pipe...

Today's optical-fiber communication systems use wavelength division multiplexing, as discussed in Section 2, to squeeze multiple channels of data through the same fiber simultaneously, offering much speedier data transfer rates.

But there is another breakthrough research in squeezing photons into this light pipe! It is the "twisted light" or the photons with an orbital angular momentum (OAM) that can also be utilized to encode data channels. It is like another degree of freedom or another dimension on which data can be transferred. The angular momentum has an infinite number of states. Each wavelength can carry different values of this angular momentum. Thus, OAM appears to be one more parameter of light that was not explored till date for communications (**Figure 11**).

Researcher Miles Padgett and his coworkers in the University of Glasgow discovered in 2004 that OAM modes can be sent through air. They used a holographic

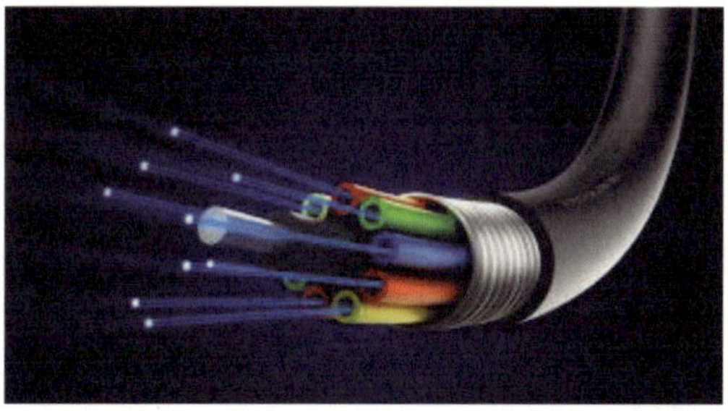

Figure 11.
Twisted light through an optical fiber (https://goo.gl/images/vAZD3C).

Figure 12.
Light trails (https://www.creativity103.com/collections/Lightwaves/slides/twisted_light_loops.html).

Figure 13.
Wireless communication through twisted light (https://www.rdmag.com/news/2017/10/twisted-light-could-illuminate-new-path-wireless-communications).

pattern to split a twisted laser light into nine separate helical beams and sent them 15 m through the air to a telescope. At the receiving end, this telescope was able to distinguish and read out all the beams simultaneously. The bandwidth of the experiment was not quite high though. A multiplexer and demultiplexer for such twisted or helical beams were presented in a research paper at the Optical Fiber Communication Conference and Exposition, in Los Angeles, in March. The multiplexing device presented in the conference was one with multiple waveguides that were carved onto a single chip. Later, Willner et al., researchers from the University of Southern California, reported a research work related to transfer of data using OAM modes of light in Nature Photonics in 2012. They had used twisted light to transfer data at approx. 2.5 terabits per second over a distance of about 1 m. But twisted beams would need to travel lot farther in order to be used for optical communications. Later, a team in Vienna, in 2014, set a record by sending pixelated images of few famous Austrians by using twisted light. The images were sent to another site in Vienna that was 3 km apart. The researchers used helical beams with four helices or twists, such that a data transfer rate of 4 pixels per second could be achieved (**Figures 12** and **13**).

The improvements should be welcome news to companies such as Intel and Luxtera, which have been racing to find ways to replace the expensive exotic semiconductors and separate components in most optical communications systems with cheap integrated chips made of silicon. Twisted light arrays could allow communication channels between chips in a computer.

5. Conclusion

Twisted light beams carrying an orbital angular momentum (OAM) have since been used to build optical tweezers, which use laser light to trap microscopic particles and control their movements. Instead of merely pushing or pulling at the particles, an OAM laser works like a tiny wrench that can torque objects around. In recent years, engineers have built ever smaller OAM beams—some barely as wide as a human blood cell—in the hopes of using them to drive microscale gears and even nanotech machiner. Biomedical diagnostic devices built on a single silicon chip could use such twisted light to operate microscopic equipment or detect the flow and viscosity of minuscule amounts of liquids. Because twisted light is so unusual, it can excite atoms and molecules into odd states not often seen in nature. Electrical engineer Natalia Litchinitser of the University at Buffalo in New York and her colleagues have used metamaterials—synthetic composites that exhibit properties not found in natural materials—to squeeze an OAM beam so that it is only a few nanometers

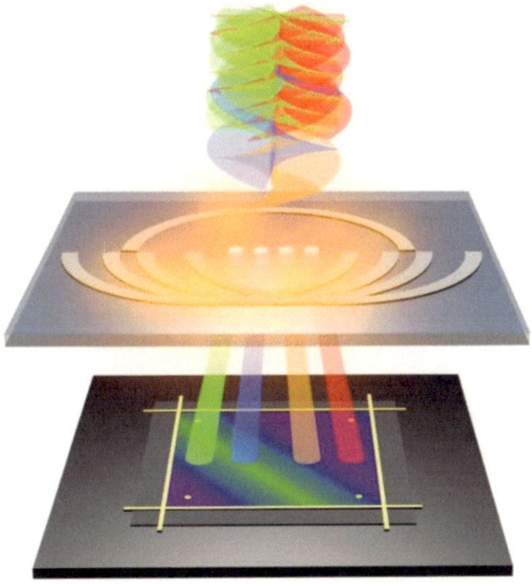

Figure 14.
The miniature OAM nano-electronic detector decodes twisted light (image courtesy: RMIT University).

Figure 15.
Twisted light sending data through an optical fiber (https://spectrum.ieee.org/tech-talk/semiconductors/design/twisted-light-sends-data-through-optical-fiber-for-first-time).

wide. Using such beams, they hope to stimulate atoms and molecules into energy states that are extremely difficult to achieve naturally. When the molecules fall back to their ground states, they release characteristic flashes of light, which Litchinitser says could be useful in new kinds of spectroscopic analysis, for instance teasing out the individual components of complex compounds (**Figure 14**).

But perhaps the biggest application of twisted light is optical communications. Recently, physicists have shown that photons are not the only ones with OAM. Pushin et al. have demonstrated that neutrons, which according to quantum mechanics act as both particles and waves, can be converted to possess OAM modes. Even acoustic waves have been induced into OAM modes, allowing them to carry more information. Some researchers have also suggested that sound waves carrying OAM can be used for underwater communication networks. These waves would travel better in water, where light is quickly absorbed otherwise (**Figure 15**).

Using twisted optical beams, one would be able to use Internet at 100 times the current speed. The growing potential of OAM beams has astounded those who work with them and surprised those who first imagined their possibility. "At one time it was just a fun idea," says Spreeuw. "I could never suspect it would grow into such an industry."

Acknowledgements

The author thanks the publisher of this book for giving her a chance to contribute a chapter.

Author details

Monika Bahl
Amity Institute of Applied Sciences, Amity University, Noida, Uttar Pradesh, India

*Address all correspondence to: monikaiitd1@gmail.com

References

[1] Allen L, Beijersbergen M, Spreeuw R, Woerdman J. Orbital angular momentum of light and the transformation of laguerre-gaussian laser modes. Physical Review A. 1992;**45**:8185-8189

[2] Barnett S. Optical angular-momentum flux. Journal of Optics B Quantum and Semiclassical Optics. 2002;**4**:S7

[3] Beijersbergen M, Allen L, van der Veen H, Woerdman J. Astigmatic laser mode converters and transfer of orbital angular momentum. Optics Communication. 1993;**96**:123-132

[4] Richard AB. Mechanical detection and measurement of the angular momentum of light. Physical Review. 1936;**50**:115-125

[5] Dholakia K, Lee W. Optical trapping takes shape: The use of structured light fields. Advances in Atomic, Molecular, and Optical Physics. 2008;**56**:261-337

[6] Dholakia K, Zemanek P. Colloquium: Gripped by light: Optical binding. Reviews of Modern Physics. 2010;**82**:1767

[7] Dholakia K, Simpson N, Padgett M, Allen L. Second-harmonic generation and the orbital angular momentum of light. Physical Review A. 1996;**54**:R3742-R3745

[8] Franke-Arnold S, Allen L, Padgett M. Advances in optical angular momentum. Laser & Photonics Reviews. 2008;**2**:299-313

[9] Gibson G, Courtial J, Padgett M, Vasnetsov M, Pas'ko V, Barnett S, et al. Freespace information transfer using light beams carrying orbital angular momentum. Optics Express. 2004;**12**:5448-5456

[10] He H, Friese M, Heckenberg N, Rubinsztein-Dunlop H. Direct observation of transfer of angular momentum to absorptive particles from a laser beam with a phase singularity. Physical Review Letters. 1995;**75**:826-829

[11] Heckenberg N, McDuff R, Smith C, White A. Generation of optical phase singularities by computer-generated holograms. Optics Letters. 1992;**17**:221-223

[12] Ladavac K, Grier D. Microoptomechanical pumps assembled and driven by holographic optical vortex arrays. Optics Express. 2004;**12**:1144-1149

[13] Leach J, Padgett M, Barnett S, Franke-Arnold S, Courtial J, Okulov A. Angular momentum of photons and phase conjugation. Physical Review Letters. 2002;**88**:257901

[14] Padgett M, Bowman R. Tweezers with a twist. Nature Photonics. 2011;**5**:343-348

[15] Parkin S, Knoner G, Nieminen T, Heckenberg N, Rubinsztein-Dunlop H. Measurement of the total optical angular momentum transfer in optical tweezers. Optics Express. 2006;**14**:6963-6970

[16] Simpson N, Dholakia K, Allen L, Padgett M. Mechanical equivalence of spin and orbital angular momentum of light: An optical spanner. Optics Letters. 1997;**22**:52-54

[17] Leach J. An optically driven pump for microfluidics. Lab on a Chip. 2006;**6**:735-739

[18] Yan Y. High-capacity millimetre-wave communications with orbital angular momentum multiplexing. Nature Communications. 2014;**5**:4876

[19] 'Twisted light' carries 2.5 terabits of data per second. BBC News. Retrieved June 25, 2012

[20] Bozinovic N. Terabit-scale orbital angular momentum mode division multiplexing in fibers. Science. June 2013;**340**:1545-1548

[21] Gregg P. Conservation of orbital angular momentum in air-core optical fibers. Optica. 2015;**2**:267-270. DOI: 10.1364/optica.2.000267

[22] Krenn M et al. Twisted light transmission over 143 kilometers. PNAS. 2016;**113**:13648-13653

[23] Agrawal G. Fiber-Optic Communication Systems. 4th ed. USA: Wiley; 2010. DOI: 10.1002/9780470918524 ISBN: 978-0-470-50511-3

[24] Gambling WA. The rise and rise of optical Fibers. IEEE Journal on Selected Topics in Quantum Electronics. 2000;**6**(6):1084-1093

[25] Mirabito MMA, Morgenstern BL. The New Communications Technologies: Applications, Policy, and Impact. 5th ed. Florida: Focal Press; 2004 ISBN: 9780-240-80586-0

[26] Mitschke F. Fiber Optics–Physics and Technology. New York: Springer; 2009 ISBN: 978-3-642-03702-3

[27] Nagel SR, MacChesney JB, Walker KL. An overview of the modified chemical vapor deposition (MCVD) process and performance. IEEE Journal of Quantum Electronics. 1982;**QE-18**(4):459

[28] Ramaswami R, Sivarajan K, Sasaki G. Optical Networks: A Practical Perspective. Burlington: Morgan Kaufmann; 2009 ISBN: 978-0-08-092072-6

[29] VDV Works LLC Lennie Lightwave's Guide To Fiber Optics, © 2002-6

Chapter 4

Cavity Generation Modeling of Fiber Fuse in Single-Mode Optical Fibers

Yoshito Shuto

Abstract

The evolution of a fiber fuse in a single-mode optical fiber was studied theoretically. To clarify both the silica-glass densification and cavity formation, which are observed in fiber fuse propagation, we investigated a nonlinear oscillation model using the Van der Pol equation. This model was able to phenomenologically explain the densification of the core material, the formation of periodic cavities, the cavity shape, and the regularity of the cavity pattern in the core layer as a result of the relaxation oscillation and cavity compression and/or deformation. Furthermore, the production and diffusion of O_2 gas in the high-temperature core layer were described on the basis of the nonlinear oscillation model.

Keywords: fiber fuse, nonlinear oscillation, Van der Pol equation

1. Introduction

Owing to the progress of dense wavelength-division multiplexing (DWDM) technology using an optical-fiber amplifier, we can exchange large amounts of data at a rate of over 100 Tbit/s over several hundred kilometers [1]. However, it is widely recognized that the maximum transmission capacity of a single strand of fiber is rapidly approaching its limit of $\sim$ 100 Tbit/s owing to the optical power limitations imposed by the fiber fuse phenomenon and the finite transmission bandwidth determined by optical-fiber amplifiers [2]. To overcome these limitations, space-division multiplexing (SDM) technology using a multicore fiber (MCF) was proposed [3, 4], and 1 Pbit/s transmission was demonstrated using a low-crosstalk 12-core fiber [5].

The fiber fuse phenomenon was first observed in 1987 by British scientists [6–9]. Several review articles [10–14] have been recently published that cover many aspects of the current understanding of fiber fuses.

A fiber fuse can be generated by bringing the end of a fiber into contact with an absorbent material or melting a small region of a fiber using an arc discharge of a fusion splice machine [6, 15–17]. If a fiber fuse is generated, an intense blue-white flash occurs in the fiber core, and this flash propagates along the core in the direction of the optical power source at a velocity on the order of 1 m/s. The temperature and pressure in the region where this flash occurs have been estimated to be about 10^4 K and 10^4 atm, respectively [18]. Fuses are terminated by gradually reducing the laser power to a termination threshold at which the energy balance in the fuse is broken.

The critical diameter d_{melted}, which is usually larger than the core diameter $2r_c$, is a characteristic dimensional parameter of the fiber fuse effect. In the inner area with diameter $d \leq d_{melted}$, a fiber fuse (high-temperature ionized gas plasma) propagates and silica glass is melted [18]. d_{melted}, defined as the diameter of the melting area, is considered as the radial size of the plasma generated in the fiber fuse [19]. Dianov *et al.* reported that the refractive index of the inner area with $d \leq d_{melted}$ in Ge-doped and/or pure silica core fibers is increased by silica-glass densification and/or the redistribution of the dopant (Ge) [20].

When a fiber fuse is generated, the core layer in which the fuse propagates is seriously damaged, and the damage has the form of periodic bullet-shaped cavities or non-periodic filaments remaining in the core [6–9, 16–32] (see **Figure 1**). Needless to say, the density in a cavity or filament is lower than that of the neighboring silica glass. It has been found that molecular oxygen is released and remains in the cavities while maintaining a high pressure (about 4 atm [7] or 5–10 atm [20]) at room temperature. Recently, several types of sensors based on periodic cavities have been proposed as a cost-effective approach to sensor production [27–29].

The dynamics of cavity formation have been investigated since the discovery of the fiber fuse phenomenon. Dianov and coworkers observed the formation of periodic bullet-shaped cavities 20–70 μs after the passage of a plasma leading edge [30, 31].

Kashyap reported that the cavity shape was dependent on the nature of the input laser light (CW or pulses) operated at a wavelength λ_0 of 1.064 μm when the average input power was maintained at 2 W [7, 15]. When CW light was input, the cavities appeared to be elliptical and cylindrically symmetric. On the other hand, short asymmetric cavities were formed by injecting (mode-locked) pulses with 100 ps FWHM (full width at half maximum), while long bullet-shaped cavities were observed by injecting pulses with 190 ps FWHM [7, 15]. Hand and Russell reported the appearance of highly regular periodic damage tracks in germanosilicate fibers at $\lambda_0 = 488$ and 514 nm [9]. Davis *et al.* reported that long non-periodic filaments occurred in germanium-doped depressed clad fibers, and a periodic damage pattern was observed in fibers doped with phosphorus and germanium at $\lambda_0 = 1.064 \mu m$ [21, 22]. Atkins *et al.* observed both periodic and long non-periodic damage tracks created in a germanosilicate-core single-mode fiber transmitting about 2 W of power at 488 nm [32]. Dianov and coworkers reported the formation of periodic damage in a silica-core fiber at 1.064 and 1.21 μm [18, 30, 31] and long non-periodic damage in a germanosilicate silica core fiber at 488 and 514 nm [20].

Todoroki classified fiber fuse propagation into three modes (unstable, unimodal, and cylindrical) according to the plasma volume relative to the pump beam size [26]. When the pump power was increased or decreased rapidly, an increase in the

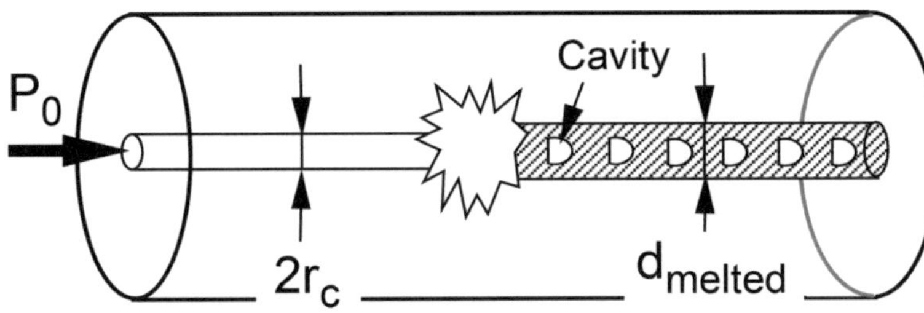

Figure 1.
Schematic view of damaged optical fiber.

length of the void-free segment or the occurrence of an irregular void pattern was observed, respectively [26].

From these observation results, the cavity patterns occurring in single-mode fibers can be classified into the four patterns shown in **Figure 2**, where l is the length of the cavity and Λ is the (periodic) cavity interval. The observed periodic cavity patterns belong to patterns (a)–(c) with the pattern depending on the value of l/Λ. The long non-periodic cavity pattern (filaments) can be considered as a sequence of two or more of pattern (d).

These cavities have been considered to be the result of either the classic Rayleigh instability caused by the capillary effect in the molten silica surrounding a vaporized fiber core [32] or the electrostatic repulsion between negatively charged layers induced at the plasma–molten silica interface [33, 34]. Although the capillary effect convincingly explains the formation mechanism of water droplets from a tap and/or bubbles through a water flow, this effect does not appear to apply to the cavity formation mechanism of a fiber fuse owing to the anomalously high viscosity of the silica glass [23, 33]. Yakovlenko proposed a novel cavity formation mechanism based on the formation of an electric charge layer on the interface between the liquid glass and plasma [33]. This charge layer, where the electrons adhere to the liquid glass surface, gives rise to a "negative" surface tension coefficient for the liquid layer. In the case of a negative surface tension coefficient, the deformation of the liquid surface proceeds, giving rise to a long bubble that is pressed into the liquid [33]. Furthermore, an increase in the charged surface due to the repulsion of similar charges results in the development of instability [33]. The instability emerges because the countercurrent flowing in the liquid causes the liquid to enter the region filled with plasma, and the extruded liquid forms a bridge. Inside the region separated from the front part of the fuse by this bridge, gas condensation and cooling of the molten silica glass occur [34]. A row of cavities is formed by the repetition of this process. Although Yakovlenko's explanation of the formation of a long cavity and rows of cavities is very interesting, the concept of "negative" surface tension appears to be unfeasible in the field of surface science and/or plasma physics (see Appendix A).

Low-frequency plasma instabilities are triggered by moving the high-temperature front of a fiber fuse toward the light source. It is well known that such a low-frequency plasma instability behaves as a Van der Pol oscillator with instability frequency ω_0 [35–55]. Therefore, the oscillatory motion of the ionized gas

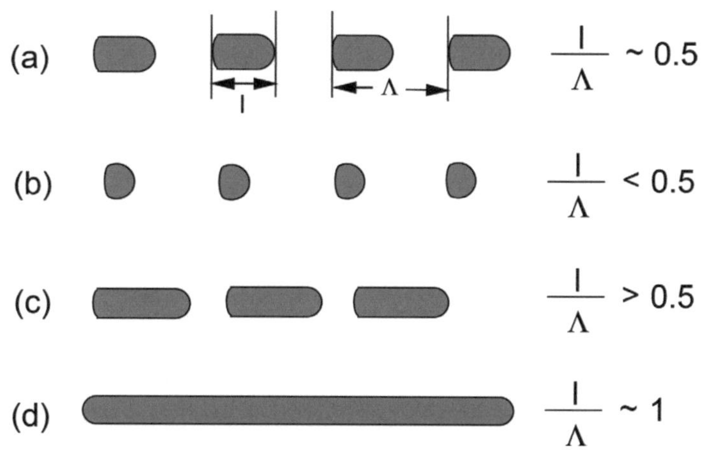

Figure 2.
Cavity patterns observed in optical fiber.

plasma during fiber fuse propagation can be studied phenomenologically using the Van der Pol equation [56].

In this paper the author describes a novel nonlinear oscillation model using the Van der Pol equation and qualitatively explains both the silica-glass densification and cavity formation observed in fiber fuse propagation. Furthermore, an investigation of the relationship between several cavity patterns and the nonlinearity parameters in the nonlinear oscillation model is reported.

2. Nonlinear oscillation behavior in ionized gas plasma

An ionized gas plasma exhibits oscillatory motion with a small amplitude when the high-temperature front of a fiber fuse propagates toward the light source.

The density ρ of the plasma is assumed to be in the form $\rho = \rho_0 + \rho_1$, where ρ_0 is the initial density of the stationary (unperturbed) part in the front region of the plasma and ρ_1 is the perturbed density. The dynamical behavior of ρ_1 resulting from fiber fuse propagation can be represented by the Van der Pol equation

$$\frac{d^2\rho_1}{dt^2} - \varepsilon\left(1 - \beta{\rho_1}^2 + 2\gamma\rho_1\right)\frac{d\rho_1}{dt} + {\omega_0}^2\rho_1 = 0, \tag{1}$$

where ε is a parameter that characterizes the degree of nonlinearity and β characterizes the nonlinear saturation (see Appendix B). The nonlinearity parameter γ characterizes the oscillation pattern.

The angular frequency ω_0 of the oscillation of the gas plasma is determined by the ion-sound velocity C_s and the free-running distance L_f of the ion-sound wave, and is given by

$$\omega_0 = 2\pi f = 2\pi\frac{C_s}{L_f}. \tag{2}$$

where f is the frequency of the oscillation of the gas plasma. The ion-sound velocity C_s is given by [38]

$$C_s = \sqrt{\frac{RT_e}{M_i}}, \tag{3}$$

where R is the gas constant, T_e is the temperature of the electron, and M_i is the mass of the ion. The author estimated $C_s = 1300$ m/s by using $T_e = 5760$ K, which was the average temperature of the radiation zone [57], and $M_i = 28 \times 10^{-3}$ kg for a Si^+ ion. The free-running distance L_f was assumed to be 1.3 mm, which was almost equal to the distance (about 1.5 mm [57]) of the radiation zone. Using Eq. (2) and the C_s (= 1300 m/s) and L_f (= 1.3 mm) values, the frequency f of the oscillation was estimated to be about 1 MHz. The relatively high f or ω_0 values reported in the literature were 426–620 kHz [52, 53] and 14.5–40.9 MHz [35, 42, 45]. These relatively high frequencies are owing to the excitation of high-frequency electron oscillation together with ion oscillation in the ionized gas plasma. The f value (= 1 MHz) estimated above is comparable to these experimental values.

The oscillatory motion for $\varepsilon = 0.1$, $\beta = 6.5$, and $\gamma = 0$ was calculated using Eq. (1). The calculated result is shown in **Figure 3**, where the perturbed density ρ_1 is plotted as a function of time. When $t \geq 80\mu s$, the maximum and minimum values of ρ_1 for the ionized gas plasma reach 0.86 and $-$ 0.86, respectively. The maximum

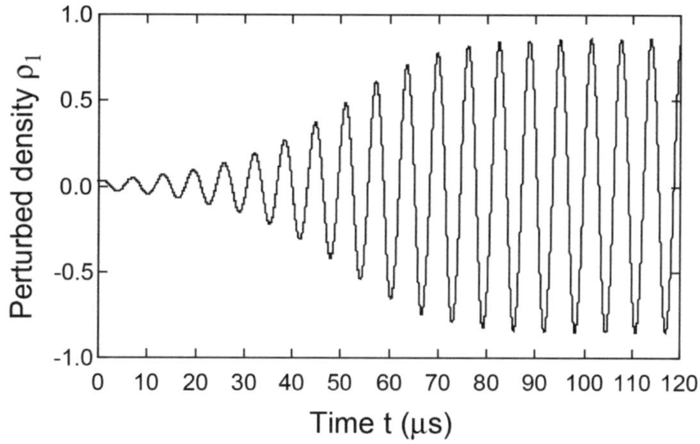

Figure 3.
Time dependence of the perturbed density during fiber fuse propagation. $\varepsilon = 0.1$, $\beta = 6.5$, $\gamma = 0$.

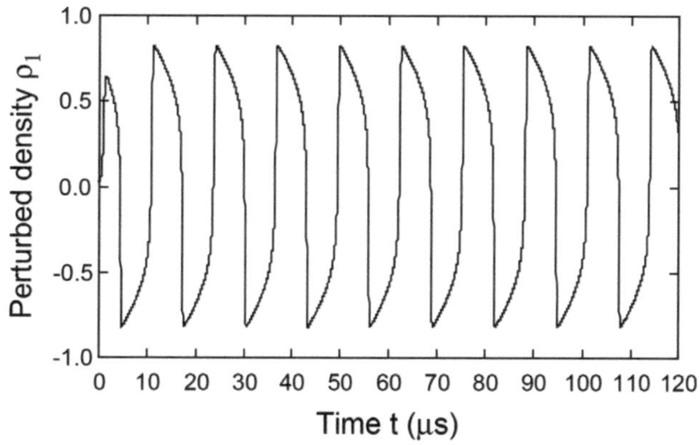

Figure 4.
Time dependence of the perturbed density during fiber fuse propagation. $\varepsilon = 5$, $\beta = 6.5$, $\gamma = 0$.

value (0.86) means that the increase in density of the core material reaches 86%, which is almost equal to the experimental value (87%) estimated by Dianov *et al.* [20].

On the other hand, it can be seen that for $\varepsilon = 0.1$ the motion of the Van der Pol oscillator is very nearly harmonic, exhibiting alternate compression and rarefaction of the density with a relatively small period Φ of about $6.3\mu s$.

Next, the oscillatory motion for $\varepsilon = 5$, 9, and 14 with $\beta = 6.5$ and $\gamma = 0$ was examined. The calculated results are shown in **Figures 4–6**, respectively. It can be seen that for $\varepsilon = 5$, 9, and 14, the oscillations consist of sudden transitions between compressed and rarefied regions. This type of motion is called a relaxation oscillation [56]. The Φ values of the motion corresponding to $\varepsilon = 5$, 9, and 14 were estimated to be about 12.9, 21.6, and 36.1 μs, respectively. These Φ values are much larger than that (about 6.3 μs) for $\varepsilon = 0.1$.

The oscillatory motion generated in the high-temperature front of the ionized gas plasma can be transmitted to the neighboring plasma at the rate of V_f when the fiber fuse propagates toward the light source. **Figure 7** shows a schematic view of the dimensional relationship between the temperature and the perturbed density of the ionized gas plasma during fiber fuse propagation.

In **Figure 7**, Λ is the interval between the periodic compressed (or rarefied) parts.

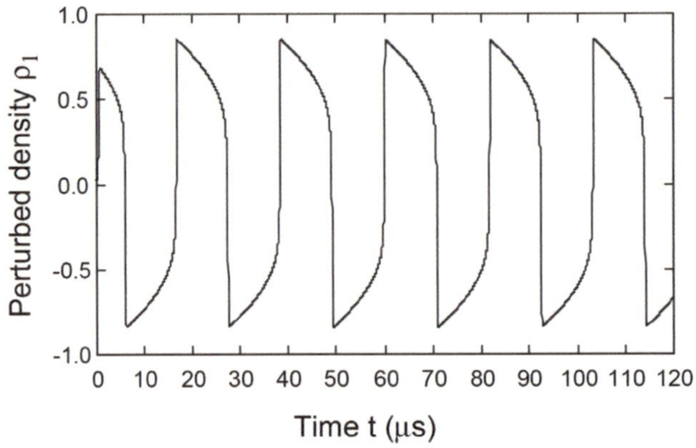

Figure 5.
Time dependence of the perturbed density during fiber fuse propagation. $\varepsilon = 9$, $\beta = 6.5$, $\gamma = 0$.

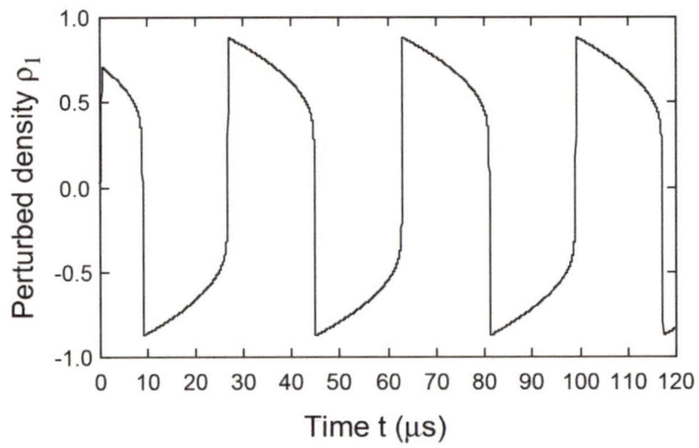

Figure 6.
Time dependence of the perturbed density during fiber fuse propagation. $\varepsilon = 14$, $\beta = 6.5$, $\gamma = 0$.

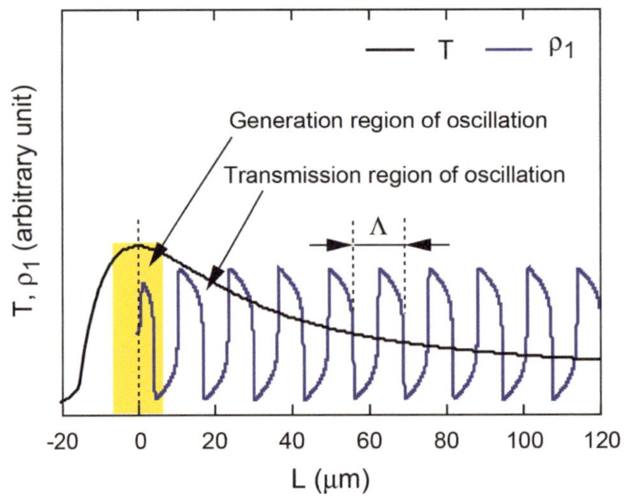

Figure 7.
Schematic view of the dimensional relationship between the temperature and the perturbed density of the ionized gas plasma during fiber fuse propagation.

The relationship between the period Φ and the interval Λ is

$$\Lambda = \Phi V_f, \tag{4}$$

where V_f is the propagation velocity of the fiber fuse and $V_f = 1$ m/s was assumed in the calculation. The Λ values of the motion corresponding to $\varepsilon = 5$, 9, and 14 are thus estimated to be about 12.9, 21.6, and 36.1 μm, respectively, using Eq. (4) and $V_f = 1$ m/s. If a large amount of molecular oxygen (O_2) accumulates in the rarefied part, the periodic formation of bubbles (or cavities) will be observed. In such a case, Λ is equal to the periodic cavity interval. The estimated Λ values (12.9, 21.6, and 36.1 μm) are close to the experimental periodic cavity intervals of 13–22 μm observed in fiber fuse propagation [13, 23].

Figure 8 shows the relationship between Φ and the nonlinearity parameter ε. As shown in **Figure 8**, Φ, which is proportional to the interval Λ, increases with increasing ε. That is, the increase in Φ and/or Λ occurs because of the enhanced nonlinearity. It was found that the experimental periodic cavity interval increases with the laser pump power [13, 23]. It can therefore be presumed that the nonlinearity of the Van der Pol oscillator occurring in the ionized gas plasma is enhanced with increasing pump power.

Kashyap reported that the cavity shape was dependent on the nature of the input laser light (CW or pulses) [7, 15]. Todoroki classified the damage to the front part of a fiber fuse into three shapes (two spheroids and a long partially cylindrical cavity) depending on the pump power [23]. He also found that a rapid increase or decrease in the pump power results in an increase in the length of the cavity-free segment or the occurrence of an irregular cavity pattern, respectively [26]. These findings indicate that the cavity shape and the regularity of the cavity pattern may be determined by the degree of nonlinearity of the Van der Pol oscillator.

In what follows, the results of examining the relationship between the interval Λ and the input laser power P_0 observed in fiber fuse propagation are described.

2.1 Power dependence of periodic cavity interval

It is well known that the fiber-fuse propagation velocity V_f increases with increasing input laser power P_0 [7, 8, 22, 23, 25, 26, 58–60]. Furthermore, in

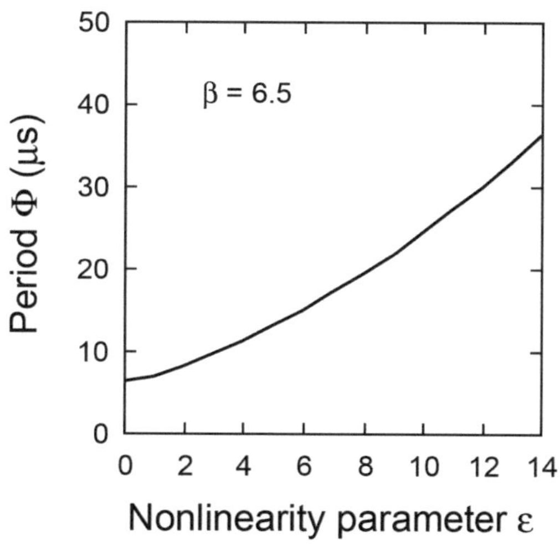

Figure 8.
Relationship between the period Φ and the nonlinearity parameter ε. $\beta = 6.5$, $\gamma = 0$.

addition to V_f, Todoroki reported the P_0 dependence of Λ in an SMF-28e fiber at $\lambda_0 = 1.48\,\mu m$ [13, 23].

In this study the author investigated the P_0 dependence of Λ using the experimental V_f values [23, 26] and the calculated Φ values shown in **Figure 8**.

To explain the experimental Λ values in the P_0 range from the threshold power ($P_{th} \simeq 1.3$W [61]) to 9 W, $\Lambda(P_0)$ can be represented by

$$\Lambda(P_0) = \Phi_0 V_f(P_0)\left[1 - \zeta\frac{\sqrt{\Phi_n(\varepsilon) - \Phi_n(\varepsilon = 0)}}{\Phi_0}\right], \tag{5}$$

where Φ_0 and ζ are constants and Φ_n is the calculated Φ value shown in **Figure 8**.

The second term $-\zeta\sqrt{\Phi_n(\varepsilon) - \Phi_n(\varepsilon = 0)}V_f(P_0)$ on the right of Eq. (5) represents the contribution of the nonlinearity to the overall Λ value.

On the other hand, the relationship between the nonlinearity parameter ε and P_0 can be expressed as

$$\varepsilon = \chi(P_0 - P_{th})^{(m/2)}, \tag{6}$$

where χ is a constant and m is the order of the square root of the power difference $P_0 - P_{th}$. ε and χ correspond to the induced polarization and nonlinear susceptibility in nonlinear optics, respectively [62]. In the calculation, the author adopted $\chi = 1$ and $m = 2$.

Using Eq. (5), $\Phi_0 = 31.5\mu s$, $\zeta = 3.6$, and the Φ_n values shown in **Figure 8**, the Λ values were calculated as a function of P_0. The calculated results are shown in **Figure 9**. The blue solid line in **Figure 9** is the curve calculated using

$$\Lambda(P_0) = \Phi_0 V_f(P_0), \tag{7}$$

which is the first term on the right of Eq. (5).

As shown in **Figure 9**, Λ increases abruptly near the threshold power (P_{th}) and increases with increasing P_0. The Λ values at $P_0 = 2.0$–2.5 W satisfy Eq. (7).

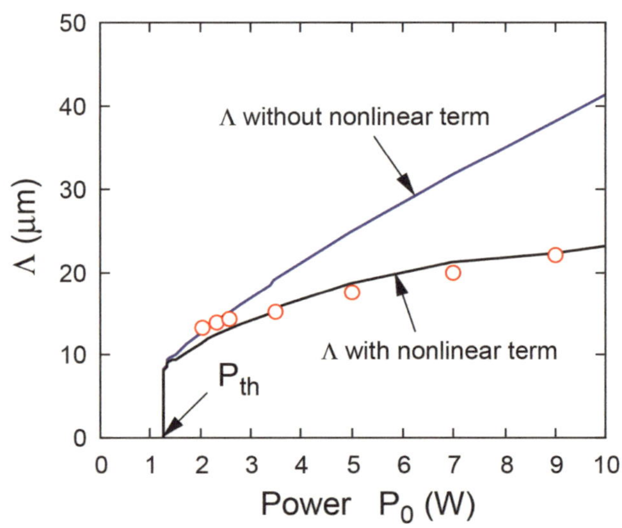

Figure 9.
Relationship between the interval Λ and the input power P_0. The blue and black solid lines were calculated using Eqs. (7) and (5), respectively. The red open circles are the data reported by Todoroki [23, 26].

However, with increasing P_0, the Λ values at P_0>2.5 W are less than those calculated using Eq. (7) and approach the Λ values estimated using Eq. (5).

This may be related to the modes of fiber fuse propagation reported by Todoroki [23, 26]. Todoroki classified the damage to the front part of a fiber fuse into three shapes (two spheroids and a long partially cylindrical cavity) depending on the pump power, and the appearance of the long partially cylindrical cavity was observed at P_0>3.5 W [23] or P_0>2.3 W [26]. As shown in **Figure 9**, the distinct contribution of the nonlinearity to the overall Λ value begins at P_0 of 2.3–3.5 W, and the oscillatory motion of the gas plasma changes from a nearly harmonic oscillation (see **Figure 3**) to a relaxation oscillation (see **Figure 4**) with increasing P_0. Therefore, the change from the spheroids of unstable and unimodal modes to the long partially cylindrical cavities of the cylindrical mode may be related to the contribution of the nonlinearity.

3. Effect of nonlinearity parameters on cavity patterns

The nonlinearity parameter γ characterizes the oscillation pattern. The oscillatory motion for $\varepsilon = 9$, $\beta = 6.5$, and $\gamma = 0$ was shown in **Figure 5**, where the perturbed density ρ_1 is plotted as a function of time. It can be seen in **Figure 5** that the oscillations consist of sudden transitions between compressed and rarefied regions, and the retention time τ_r of the rarefied regions equals that of the compressed regions τ_c. The relationship between the period Φ $(= \tau_r + \tau_c)$ and the interval Λ is given by Eq. (4), and the relationship between τ_r and the length l of the cavity is

$$l = \tau_r V_f. \tag{8}$$

The Λ and l values of the motion corresponding to $\varepsilon = 9$, $\beta = 6.5$, and $\gamma = 0$ are estimated to be about 10.8 and 21.6 μm, respectively, using Eqs. (4) and (8) and $V_f = 1$ m/s. That is, $l/\Lambda = 0.5$ in the case of $\gamma = 0$.

Next, the oscillatory motion for $\gamma = 2$ and -2 with $\varepsilon = 9$ and $\beta = 6.5$ was examined. The calculated results are shown in **Figures 10** and **11**, respectively. As shown in **Figure 10**, the retention time τ_r of the rarefied regions is larger than that of the compressed regions τ_c. As a result, the ratio l/Λ is larger than 0.5 in the case of

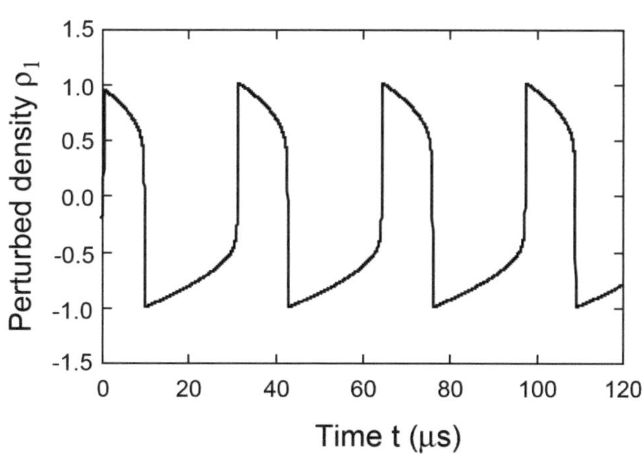

Figure 10.
Time dependence of the perturbed density during fiber fuse propagation. $\varepsilon = 9$, $\beta = 6.5$, $\gamma = 2$.

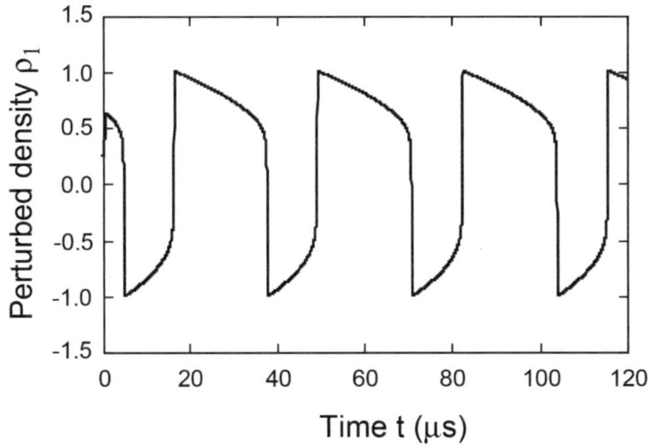

Figure 11.
Time dependence of the perturbed density during fiber fuse propagation. $\varepsilon = 9$, $\beta = 6.5$, $\gamma = -2$.

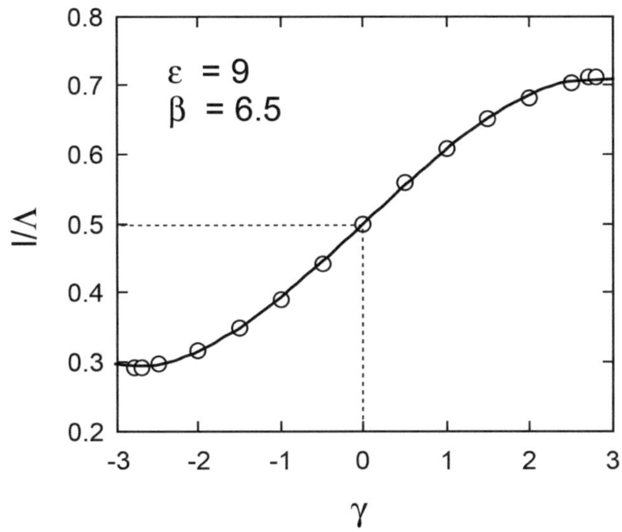

Figure 12.
Relationship between l/Λ *and the nonlinearity parameter* γ. $\varepsilon = 9$, $\beta = 6.5$.

$\gamma = 2$. On the other hand, as shown in **Figure 11**, τ_r is smaller than τ_c and $l/\Lambda < 0.5$ in the case of $\gamma = -2$.

Figure 12 shows the relationship between l/Λ and the nonlinearity parameter γ. As shown in **Figure 12**, l/Λ increases with increasing γ and approaches its maximum value (about 0.71) at $\gamma \sim 2.8$. In contrast, l/Λ approaches its minimum value (about 0.29) at $\gamma \sim -2.8$.

3.1 Deformation of cladding due to plasma formation

The inside of the high-temperature core of 4,000–10,000 K has a high internal pressure p of 1×10^4–5×10^4 atm [18]. The inner wall of the core (in the solid state) will be expanded by this internal pressure p. To simplify the calculation, the existence of molten silica glass (liquid state) between the solid-state cladding layer (inner radius r_i, outer radius r_f) and the inner high-pressure gas plasma is ignored [33].

r_i for the cladding is assumed to be $d_{melted}/2$. With increasing inner pressure p, the inner radius of the cladding layer increases in the radial direction owing to the compression of the cladding layer. The increment δr in the radius r of the solid-state cladding layer can be expressed in terms of the Young's modulus E and Poisson's ratio ν of the (solid-state) silica glass, and is given by the following equation [63].

$$\delta r = \frac{r_i^2 p}{E\left(r_f^2 - r_i^2\right)} \left[(1-\nu) + (1+\nu)\frac{r_f^2}{r^2}\right] \cdot r \tag{9}$$

Todoroki reported that d_{melted} and the diameter d of periodic cavities with $\Lambda \sim 22\mu m$, which is equal to that in the case of $\varepsilon = 9$ and $\gamma = 0$, were about 20 and 6.5 μm, respectively [13]. We adopted $r_i = d_{melted}/2 \cong 10\,\mu m$ and $r_f = 62.5\mu m$. Using $E = 73$ GPa and Poisson's ratio $\nu = 0.17$ for silica glass, the relationship between $\delta r/r_i$ and r/r_i at $p = 2$ GPa (=1.97 × 10^4 atm) is calculated. The results are shown in **Figure 13**. It can be clearly seen from **Figure 13** that the elongation rate $\delta r/r_i$ of the inner radius has a maximum value (about 3.35%) when r/r_i.

We consider the tensile stress σ_θ acting on the inner wall ($r = r_i$) of the cladding layer. σ_θ is related to p by the following expression [63]:

$$\sigma_\theta = \frac{r_f^2 + r_i^2}{r_f^2 - r_i^2} \cdot p. \tag{10}$$

σ_θ increases with increasing p. Using $r_i \sim 10\,\mu m$ and $r_f = 62.5\mu m$, σ_θ was estimated to be about 2.1 GPa when $p = 2$ GPa. If this σ_θ value exceeds the ideal fracture strength σ_0 of the silica glass, a crack will be generated on the inner wall of the cladding layer.

On the other hand, it is well known for various solid materials that the σ_0 value is related to the Young's modulus E of the material by the following equation [64]:

$$\sigma_0 \approx E/10. \tag{11}$$

By using Eq. (11) and $E = 73$ GPa for silica glass, we can estimate σ_0 to be approximately 7.3 GPa. Since this value is larger than the estimated σ_θ value (2.1 GPa), the cladding layer is never broken, but it can be seen that a relatively large expansion of the inner radius occurs as a result of the internal pressure.

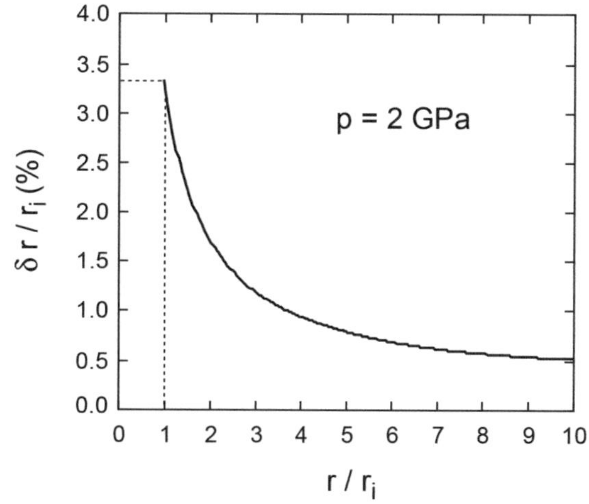

Figure 13.
Relationship between $\delta r/r_i$ and r/r_i.

The excess volume ΔV produced by the expansion of the inner radius over the interval Λ of the cavity can be estimated as follows using the maximum δr value δr_{max} at $r = r_i$:

$$\Delta V = \Lambda\pi\left[(r_i + \delta r_{max})^2 - r_i^2\right]. \tag{12}$$

As the maximum elongation rate $\delta r_{max}/r_i$ was about 3.35% (see **Figure 13**), δr_{max} was estimated to be about 0.335 μm by using $r_i \cong 10\,\mu m$.

On the other hand, the volume V of a cavity with diameter d and length l is given by

$$V = l\pi\left(\frac{d}{2}\right)^2. \tag{13}$$

It is considered that the volume required to generate a cavity was compensated by the excess volume ΔV [33]. If the value of V required to generate a cavity in the interval Λ is smaller than ΔV, the oscillation pattern predicted by Eq. (1) will be maintained and periodic cavities having a size corresponding to V will be formed in the core. That is, the necessary condition for the formation of a periodic cavity pattern is that the ratio of V to ΔV is smaller than 1, which is expressed as follows:

$$\frac{V}{\Delta V} = \frac{l}{\Lambda}\frac{d^2}{4\delta r_{max}(2r_i + \delta r_{max})} \leq 1. \tag{14}$$

Rearranging Eq. (14), we obtain the following inequality for l/Λ:

$$\frac{l}{\Lambda} \leq \frac{4}{d^2}\delta r_{max}(2r_i + \delta r_{max}). \tag{15}$$

When $r_i \sim 10\,\mu m$, $\delta r_{max} \sim 0.335\mu m$, and $d \sim 6.5\mu m$, we obtain

$$\frac{l}{\Lambda} \leq 0.645.$$

When l/Λ satisfies this condition, the periodic cavities predicted by Eq. (1) will be formed in the core.

However, as shown in **Figure 12**, l/Λ can be larger than 0.645 when $\gamma > 1.5$. In this case, the cavities formed in the core will be compressed and deformed as shown in **Figure 14**.

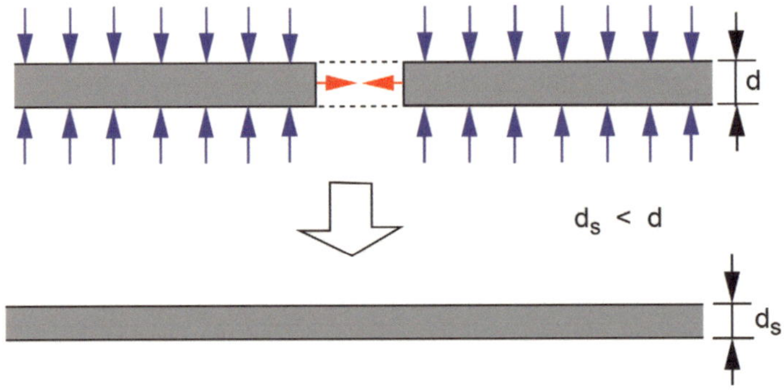

Figure 14.
Schematic view of cavity compression and deformation in core.

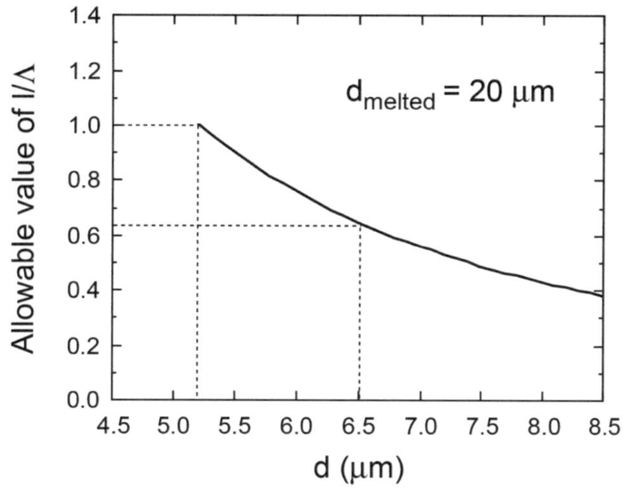

Figure 15.
Relationship between the maximum allowable value of l/Λ *and the cavity diameter d.* $d_{melted} = 20\ \mu m$.

As shown in Eq. (15), the allowable value of l/Λ increases with decreasing cavity diameter d. **Figure 15** shows the relationship between the maximum allowable value of l/Λ and the diameter d. As shown in **Figure 15**, when d is reduced by 20% from 6.5 to 5.2 μm, we obtain.

$$\frac{l}{\Lambda} \leq 1.$$

Under this condition, cavity pattern (d) (long filaments) in addition to periodic pattern (c) in **Figure 2** can be formed in the core. As the number of repetitions of pattern (d) can change freely, the period of long filaments can be irregular. This may be the cause of the long non-periodic filaments observed by several researchers [20–22, 32].

Kashyap reported that the diameter of a short asymmetric cavity with $l/\Lambda < 0.5$ was larger than that of an oblong and cylindrically symmetric cavity with l/Λ of about 0.5 and that the diameter of a long bullet-shaped cavity with $l/\Lambda < 0.5$ was smaller than that of the cavities described above [7]. These findings are consistent with the calculation results shown in **Figure 15**. In what follows, the production and diffusion of O_2 gas in the high-temperature core layer are described.

3.2 Oxygen production in optical Fiber

When gaseous SiO and/or SiO_2 molecules are heated to high temperatures of above 5,000 K, they decompose to form Si and O atoms, and finally become Si^+ and O^+ ions and electrons in the ionized gas plasma state.

In a confined core zone, and thus at high pressures, SiO_2 is decomposed with the evolution of SiO gas or Si and O atomic gases at elevated temperatures [65]:

$$SiO_2 \rightleftarrows SiO + (1/2)\,O_2 \rightleftarrows Si + 2O. \tag{16}$$

The number densities N_{SiO}, N_{Si}, and N_O (in cm^{-3}) can be estimated using the procedure described in [57, 66] and the published thermochemical data [67] for Si, SiO, O, O_2, and SiO_2.

The dependence of N_O on the temperature T is shown in **Figure 16**. N_O gradually approaches its maximum value ($3.3 \times 10^{21} cm^{-3}$) at 11,100 K and then decreases

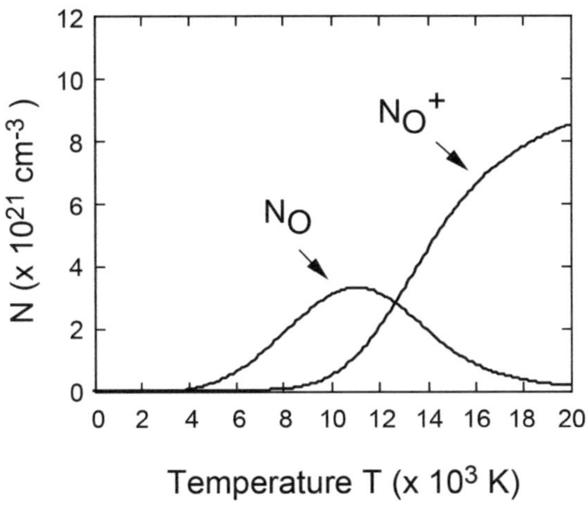

Figure 16.
Temperature dependences of the number densities of O and O^+.

with further increasing T. This is because oxygen (O) atoms are ionized to produce O^+ ions and electrons in the ionized gas plasma as follows:

$$O \rightleftarrows O^+ + e^-. \tag{17}$$

The number density N_{O^+} of O^+ ions can be estimated using the Saha equation [66, 68]:

$$\frac{N_{O^+}^2}{N_O} \approx 2 \frac{(2\pi m_e kT)^{3/2}}{h^3} \frac{Z_+}{Z_0} \exp\left(-I_p/k_B T\right), \tag{18}$$

where I_p (= 13.61 eV [69]) is the ionization energy of a neutral O atom, m_e is the electron mass, h is Planck's constant, and k_B is Boltzmann's constant. Z_+ and Z_0 are the partition functions of ionized atoms and neutral atoms, respectively, and $Z_+ \approx Z_0$. The relationship between N_{O^+} and T is also shown in **Figure 16**. N_{O^+} increases gradually at temperatures above 7,000 K and reaches $8.9 \times 10^{21}\,\mathrm{cm}^{-3}$ at 2×10^4 K.

It has been found that molecular oxygen is released and remains in the cavities of a damaged core layer while maintaining a relatively high pressure (about 4 atm [7] or 5–10 atm [20]) at room temperature. The molecular oxygen (O_2) is produced from neutral O atoms as follows:

$$2O \rightarrow O_2. \tag{19}$$

The rate equation of this reaction is [70]

$$\frac{dN_{O_2}}{dt} = \sqrt{2}\pi\sigma^2 \sqrt{\frac{8RT}{\pi M_O}} {N_O}^2 \exp\left(-E_a/RT\right), \tag{20}$$

where σ (= 1.5 Å) is half of the collision diameter, M_O ($= 16.0 \times 10^{-3}$ kg) is the atomic weight of O, and E_a is the activation energy. The bond energy (493.6 kJ/mol [71]) of oxygen was used for E_a.

The dependence of dN_{O_2}/dt on the temperature T is shown in **Figure 17**. The rate of O_2 production dN_{O_2}/dt exhibits its maximum value ($2.96 \times 10^{31}\,\mathrm{cm}^{-3}\mathrm{s}^{-1}$) at 12,700 K. This means that the oxygen molecules are produced most effectively at 12,700 K.

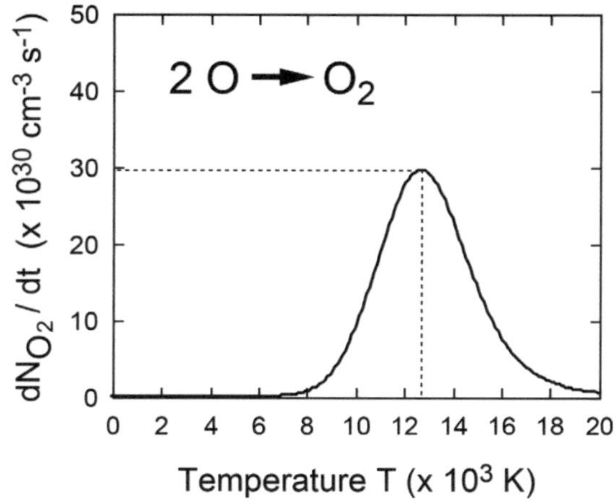

Figure 17.
Temperature dependence of the production rate of O_2.

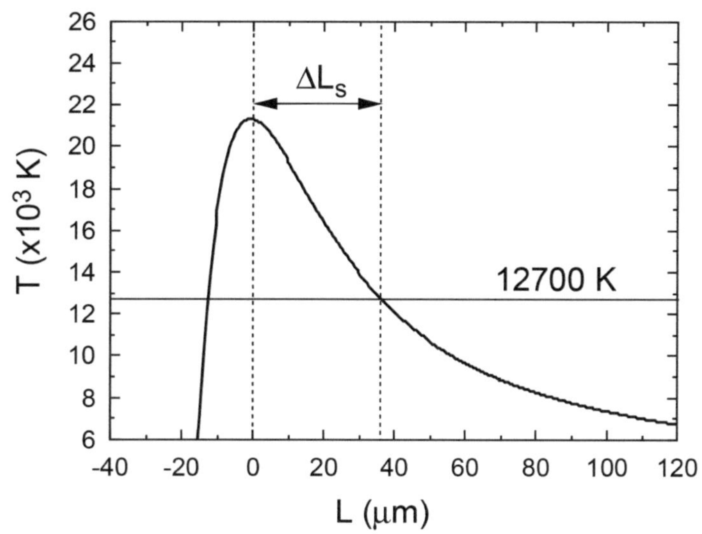

Figure 18.
Temperature distribution of the high-temperature front versus the length along the z direction at $t = 3$ *ms after the incidence of 1.8 W laser light for* $IA = 8$ *dB. The center of the high-temperature front is set at* $L = 0$ *μm.*

Figure 18 shows the temperature distribution of the high-temperature front along the z direction at $t = 3$ ms after the incidence of 1.8 W laser light for $IA = 8$ dB. The calculation of the temperature distribution was described in Ref. [72]. In **Figure 18** the initial attenuation IA of 8 dB corresponds to an optical absorption coefficient α of $1.84 \times 10^6\,\mathrm{m}^{-1}$ when the thickness of the absorption layer, which consists of carbon black, is about 1 μm [72]. In this figure, the center of the high-temperature front is set at $L = 0$ μm. As shown in **Figure 18**, ΔL_s, which is about 36.5 μm, is the distance between the high-temperature peak ($L = 0$ μm) and the location with a temperature of 12,700 K.

This ΔL_s can be converted into the time lag $\Delta \tau_s$ from the passage of the high-temperature front as follows:

$$\Delta \tau_s = \frac{\Delta L_s}{V_f}. \tag{21}$$

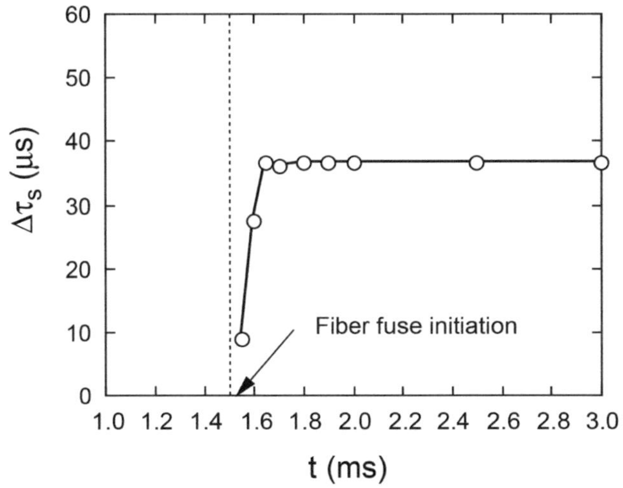

Figure 19.
$\Delta\tau_s$ values versus t after the incidence of 1.8 W laser light for IA = 8 dB.

It is expected that the O_2 molecular gas in the ionized gas plasma will be observed most frequently after a time lag of $\Delta\tau_s$ from the passage of the high-temperature peak. If the produced O_2 gas diffuses into the rarefied part of the oscillatory variation in density shown in **Figures 4–6**, **10**, and **11**, periodic cavities containing some of the oxygen molecules will be formed (see below).

When $V_f = 1$ m/s, the $\Delta\tau_s$ values were estimated at a time of $t = 1.55$–3 ms after the incidence of 1.8 W laser light for $IA = 8$ dB. The calculated $\Delta\tau_s$ values are plotted in **Figure 19** as a function of t. The fiber fuse phenomenon was initiated at $t = 1.5$ ms (see **Figure 14** in Ref. [72]). As shown in **Figure 19**, $\Delta\tau_s$ increases rapidly with increasing t immediately after the fiber fuse is initiated and reaches a constant value (36.5 μs) at $t>$ 1.65 ms. This value is in reasonable agreement with the experimental values (20–70 μs) reported by Dianov and coworkers [30, 31].

3.3 Diffusion length of oxygen gas

The O_2 gas produced near the high-temperature front diffuses from the compressed part into the rarefied part of the oscillatory variation during a short period Φ of 10–30 μs (see **Figure 8**).

The diffusion coefficient D of the O_2 gas is given by [70].

$$D = \frac{2}{3\pi\sigma^2 N_{O_2}}\sqrt{\frac{RT}{\pi M_{O_2}}}, \tag{22}$$

where M_{O_2} ($= 32.0 \times 10^{-3}$ kg) is the molecular weight of O_2 gas. As N_{O_2} is smaller than $N_O/2$, $N_{O_2} \approx N_O/2$ is assumed in the calculation.

The mean square of the displacement $\Delta\bar{z}^2$ along the z direction of the optical fiber can be estimated from D and time t as follows [73]:

$$\Delta\bar{z}^2 = 2Dt. \tag{23}$$

The $\Delta\bar{z}$ values at $T = 12,700$ K were estimated using Eqs. (16) and (17). When $t = 20\,\mu s$, the calculated $\Delta\bar{z}$ value is given by

$$\Delta\bar{z} = \pm 16.7\ \mu m.$$

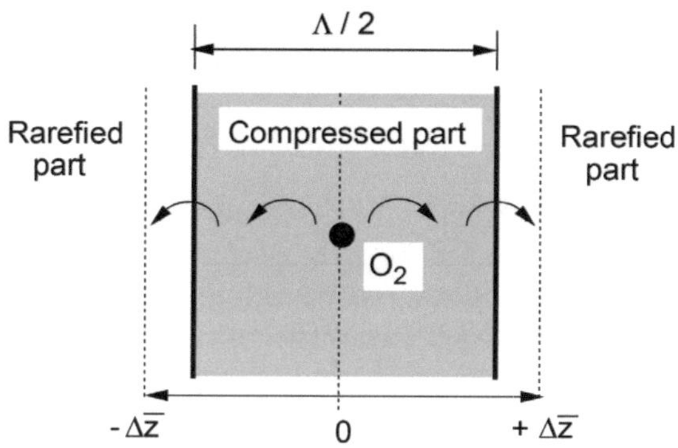

Figure 20.
Schematic view of diffusion of oxygen gas from the compressed part into the rarefied part in the high-temperature plasma.

This $\Delta\bar{z}$ value is of the same order as the observed periodic cavity interval (13–22 μm) [13].

Figure 20 shows a schematic view of the diffusion of the O_2 gas from the compressed part into the rarefied part in the high-temperature plasma. If the absolute value of $\Delta\bar{z}$ is larger than half of the interval Λ between the periodic rarefied parts, many of the O_2 molecules produced in the compressed part can move into the rarefied part during the period Φ (10–30 μs) of the relaxation oscillation. This O_2 gas will form temporary microscopic cavities that can constitute the nuclei necessary for growth into macroscopic bubbles [74].

As described above, the nonlinear oscillation model was able to phenomenologically explain both the densification of the core material and the formation of periodic cavities in the core layer as a result of the relaxation oscillation and the formation of O_2 gas near the high-temperature front.

4. Conclusion

The evolution of a fiber fuse in a single-mode optical fiber was studied theoretically. To clarify both the silica-glass densification and cavity formation, which are observed in fiber fuse propagation, we investigated a nonlinear oscillation model using the Van der Pol equation. This model was able to phenomenologically explain the densification of the core material, the formation of periodic cavities, the cavity shape, and the regularity of the cavity pattern in the core layer as a result of the relaxation oscillation and cavity compression and/or deformation.

This nonlinear oscillation model including the relaxation oscillation is a phenomenological model, and the relationship between the nonlinearity parameters (ε, β, γ) and the physical properties observed in the fiber fuse experiments is unknown. Therefore, to clarify this relationship, further quantitative investigation is necessary.

A. Electrostatic interaction between charged surface and plasma

In a confined core zone, and thus at a high pressure, SiO_2 is decomposed with the evolution of SiO gas or Si and O atomic gases at elevated temperatures, as described in the main text. When the Si and O atomic gases are heated to high temperatures of above 3,000 K (Si) and 4,000 K (O), they are ionized to produce Si^+ and O^+ ions and electrons in the ionized gas plasma state.

$$Si + O \rightleftarrows Si^+ + O^+ + 2e^- \tag{24}$$

If thermally produced electrons in the plasma are not bound to positive species (Si^+ or O^+ ions), they can move freely in the plasma under the action of the alternating electric field of the light wave. Such free diffusion is possible only in the limiting case of very low charge densities. However, as shown in **Figure 16** and also **Figure 1** in Ref. [66], the densities of Si^+ and O^+ ions and electrons are reasonably large above 1×10^4 K. At high charge densities, it is known that the positive and negative species diffuse at the same rate. This phenomenon, proposed by Schottky [75], is called ambipolar diffusion [76, 77]. Ambipolar diffusion is the diffusion of positive and negative species owing to their interaction via an electric field (space-charge field). In plasma physics, ambipolar diffusion is closely related to the concept of quasineutrality.

Some electrons arrive at the surface of melted silica glass, and they attach to oxygen atoms on the surface because oxygen atoms have a high electron affinity [78]. As a result, a negatively charged surface, which was proposed by Yakovlenko [33], may be formed as shown in **Figure 21**.

However, the negative charges on the surface will immediately be balanced by an equal number of oppositely charged Si^+ and O^+ ions because these positive ions move together with the electrons as a result of ambipolar diffusion. In this way, an atmosphere of ions is formed in the rapid thermal motion close to the surface. This ionic atmosphere is known as the diffuse electric double layer [79].

The thickness δ_0 of the double layer is approximately $1/\kappa$, which is the characteristic length known as the Debye length. The parameter κ is given in terms of N_e and T as follows [77]:

$$\kappa^2 = \frac{2N_e e^2}{\varepsilon_0 k_B T}, \tag{25}$$

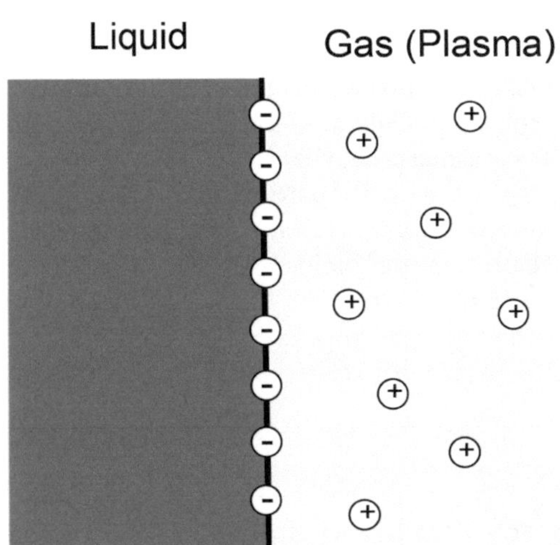

Figure 21.
Schematic view of the negatively charged surface and ionic atmosphere.

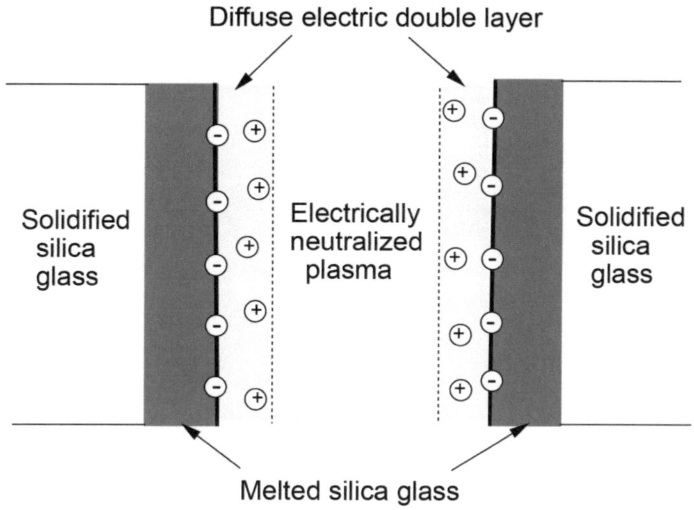

Figure 22.
Schematic view of the cross section of the high-temperature plasma in the optical fiber.

where e is the charge of an electron and ε_0 is the dielectric constant of vacuum. When $T = 1 \times 10^4$ K, $N_e = 2.2 \times 10^{20}$ cm^{-3}. Using these values and Eq. (25), the thickness δ_0 of the double layer at 1×10^4 K was estimated to be about 3.3×10^{-10} m.

A cross section of the high-temperature plasma in the optical fiber with the double layers is schematically shown in **Figure 22.**

In the central domain of the high-temperature plasma, electrically neutral atoms (Si and O) and charged species (Si^+, O^+, and e^-) exist. As the charged species are balanced, electrical neutrality is achieved in the domain. Moreover, the dimensions of the domain are almost equal to those of the high-temperature plasma excluding the very thin (Å order) electric double layers at the surface of the melted silica glass.

B. Nonlinearity parameter β in Van der pol equation

The dynamical behavior of the perturbed density ρ_1 resulting from fiber fuse propagation can be represented by the Van der Pol equation

$$\ddot{\rho}_1 - \varepsilon\left(1 - \beta{\rho_1}^2\right)\dot{\rho}_1 + {\omega_0}^2\rho_1 = 0, \tag{26}$$

where $\ddot{\rho}_1 = d^2\rho_1/dt^2$, $\dot{\rho}_1 = d\rho_1/dt$, ε and β are nonlinearity parameters, and the nonlinearity parameter $\gamma = 0$ is assumed.

If the solution of Eq. (26) is written as

$$\rho_1 = A\cos\left(\omega_0 t + \varphi\right), \tag{27}$$

where the amplitude A and phase φ are slowly varying functions, then A satisfies the following equation:

$$A^2 = {\rho_1}^2 + \left(\frac{\dot{\rho}_1}{\omega_0}\right)^2. \tag{28}$$

Differentiating Eq. (28), we obtain

$$\begin{aligned}\dot{A} &= \frac{\dot{\rho}_1}{\omega_0{}^2 A}\left(\ddot{\rho}_1 + \omega_0{}^2\rho_1\right)\\ &= \frac{\dot{\rho}_1}{\omega_0{}^2 A}\left[\varepsilon\left(1-\beta{\rho_1}^2\right)\dot{\rho}_1\right]\\ &= \frac{\varepsilon}{\omega_0{}^2 A}\left(\dot{\rho}_1{}^2\right) - \frac{\varepsilon\beta}{\omega_0{}^2 A}{\rho_1}^2\left(\dot{\rho}_1{}^2\right)\\ &= \varepsilon A\sin^2(\omega_0 t+\varphi) - \varepsilon\beta A^3\sin^2(\omega_0 t+\varphi)\cos^2(\omega_0 t+\varphi)\\ &= \frac{\varepsilon}{2}A[1-\cos(2\omega_0 t+2\varphi)] - \frac{\varepsilon\beta}{8}A^3[1-\cos(4\omega_0 t+4\varphi)].\end{aligned} \tag{29}$$

Because of the slowly varying property of A, the oscillatory terms $A\cos(2\omega_0 t+2\varphi)$ and $A^3\cos(4\omega_0 t+4\varphi)$ on the right of Eq. (29) are averaged out every cycle and can be discarded [80], thus reducing Eq. (29) to

$$\begin{aligned}\dot{A} &\simeq \frac{\varepsilon}{2}A - \frac{\varepsilon\beta}{8}A^3\\ &\simeq \frac{\varepsilon}{2}A\left(1-\frac{\beta}{4}A^2\right).\end{aligned} \tag{30}$$

The maximum value of A, A_m, is obtained under the condition of $\dot{A}=0$. To satisfy this condition,

$$A_m = \frac{2}{\sqrt{\beta}}. \tag{31}$$

This means that the nonlinearity parameter β determines the maximum and minimum values of ρ_1. In the calculation, we used $\beta = 6.5$, which corresponds to $A_m \simeq 0.8$.

Author details

Yoshito Shuto
Ofra Project, Iruma, Japan

*Address all correspondence to: ofra@tuba.ocn.ne.jp

References

[1] Sano A, Kobayashi T, Yamanaka S, Matsuura A, Kawakami H, Miyamoto Y, et al. 102.3-Tb/s (224 x 548-Gb/s) C- and extended L-band all-Raman transmission over 240 km using PDM-64QAM single carrier FDM with digital pilot tone. Optical Fiber Communication Conference Exhibition/ National Fiber Optic Engineers Conference 2012 (OFC/NFOEC2012); PDP5C.3

[2] Nakazawa M. Evolution of EDFA from single-core to multi-core and related recent progress in optical communication. Optical Review. 2014; **21**(6):862-874

[3] Morioka T. New generation optical infrastructure technologies: "EXAT initiative" toward 2020 and beyond. OptoElectronics and Communications Conference 2009 (OECC 2009); FT4

[4] Richardson DJ, Fini JM, Nelson LE. Space-division multiplexing in optical fibres. Nature Photonics. 2013;7:354-362

[5] Takara H, Asano A, Kobayashi T, Kubota H, Kawakami H, Matsuura A, et al. 1.01-Pb/s (12 SDM/222 WDM/456 Gb/s) crosstalk-managed transmission with 91.4-b/s/Hz aggregate spectral efficiency. European Conference on Optical Communication 2012 (ECOC2012); Th.3.C.1

[6] Kashyap R, Blow KJ. Observation of catastrophic self-propelled self-focusing in optical fibres. Electronics Letters. 1988;**24**(1):47-49

[7] Kashyap R. Self-propelled self-focusing damage in optical fibres. Proceedings of Xth International Conference on Lasers; 1988. pp. 859-866

[8] Hand DP, Russell PSJ. Solitary thermal shock waves and optical damage in optical fibers: The fiber fuse. Optics Letters. 1988;**13**(9):767-769

[9] Hand DP, Russell PSJ. Soliton-like thermal shock-waves in optical fibers: Origin of periodic damage tracks. European Conference on Optical Communications. 1988:111-114

[10] André P, Rocha A, Domingues F, Facão M. Thermal effects in optical fibres. In: Bernardes MAD, editor. Developments in Heat Transfer. Rijeka, Croatia: InTech; 2011

[11] Todoroki S. Fiber fuse propagation behavior. In: Moh Y, Harun SW, Arof H, editors. Selected Topics on Optical Fiber Technology. Rijeka, Croatia: InTech; 2012

[12] Kashyap R. The fiber fuse–From a curious effect to a critical issue: A 25th year retrospective. Optics Express. 2013; **21**(5):6422-6441

[13] Todoroki S. Fiber Fuse: Light-Induced Continuous Breakdown of Silica Glass Optical Fiber. NIMS Monographs. Tokyo: Springer; 2014

[14] Shuto Y. Simulation of fiber fuse phenomenon in single-mode optical fibers. In: Yasin M, Arof H, Harun SW, editors. Advances in Optical Fiber Technology. Rijeka, Croatia: InTech; 2014

[15] Kashyap R. High average power effects in optical fibres and devices. In: Limberger HG, Matthewson MJ, editors. Proceedings of Society of Photo-Optical Instrumentation Engineers. 2003. Vol. 4940. pp. 108-117

[16] Todoroki S. Quantitative evaluation of fiber fuse initiation probability in typical single-mode fibers. Optical Fiber Communication Conference; 2015 (OFC2015); W2A.33

[17] Todoroki S. Quantitative evaluation of fiber fuse initiation with exposure to arc discharge provided by a fusion

splicer. Scientific Reports. 2016;**6**: 25366-1-25366-6

[18] Dianov EM, Bufetov IA, Frolov AA. Destruction of silica fiber cladding by the fuse effect. Optics Letters. 2004; **29**(16):1852-1854

[19] Kurokawa K, Hanzawa N. Fiber fuse propagation and its suppression in hole assisted fibers. IEICE Transactions on Communications. 2011;**E94-B**(2): 384-391

[20] Dianov EM, Mashinsky VM, Myzina VA, Sidorin YS, Streltsov AM, Chickolini AV. Change of refractive index profile in the process of laser-induced fibre damage. Soviet Lightwave Commun. 1992;**2**:293-299

[21] Davis DD, Mettler SC, DiGiovanni DJ. Experimental data on the fiber fuse. In: Bennett HE, Guenther AH, Kozlowski MR, Newnam BE, Soileau MJ, editors. Proceedings of Society of Photo-Optical Instrumental Engineers. 1995;**2714**:202-210

[22] Davis DD, Mettler SC, DiGiovanni DJ. A comparative evaluation of fiber fuse models. In: Bennett HE, Guenther AH, Kozlowski MR, Newnam BE, Soileau MJ, editors. Proceedings of Society of Photo-Optical Instrumental Engineers. 1966;**2966**:592-606

[23] Todoroki S. Origin of periodic void formation during fiber fuse. Optics Express. 2005;**13**(17):6381-6389

[24] Todoroki S. In situ observation of modulated light emission of fiber fuse synchronized with void train over hetero-core splice point. PLoS One. 2008;**3**(9):e3276-1–4

[25] Domingues F, Frias AR, Antunes P, Sousa AOP, Ferreira RAS, André PS. Observation of fuse effect discharge zone nonlinear velocity regime in erbium-doped fibres. Electronics Letters. 2012;**48**(20):1295-1296

[26] S. Todoroki. Fiber fuse propagation modes in typical single-mode fibers. Optical Fiber Communication Conference 2013 (OFC2013); JW2A.11

[27] Antunes PFC, Domingues MFF, Alberto NJ, André PS. Optical fiber microcavity strain sensors produced by the catastrophic fuse effect. IEEE Photonics Technology Letters. 2014; **26**(1):78-81

[28] Lin G, Baiad MD, Gagne M, Liu W, Kashyap R. Harnessing the fiber fuse for sensing applications. Optics Express. 2014;**22**(8):8962-8969

[29] Domingues MFF, Paixão TB, Mesquita EFT, Alberto N, Frias AR, Ferreira RAS, et al. Liquid hydrostatic pressure optical sensor based on micro-cavity produced by the catastrophic fuse effect. IEEE Sensors Journal. 2015; **15**(10):5654-5658

[30] Bufetov IA, Frolov AA, Dianov EM, Frotov VE, Efremov VP. Dynamics of fiber fuse propagation. Optical Fiber Communication Conference Exhibition/ National Fiber Optical Engineering Conference 2005 (OFC/NFOEC2005); OThQ7

[31] Dianov EM, Frotov VE, Bufetov IA, Efremov VP, Rakitin AE, Melkumov MA, et al. High-speed photography, spectra, and temperature of optical discharge in silica-based fibers. IEEE Photonics Technology Letters. 2006; **18**(6):752-754

[32] Atkins RM, Simpkins PG, Yabon AD. Track of a fiber fuse: A Rayleigh instability in optical waveguides. Optics Letters. 2003;**28**(12):974-976

[33] Yakovlenko SI. Plasma behind the front of a damage wave and the mechanism of laser-induced production of a chain of caverns in an optical fibre. Quantum Electronics. 2004;**34**(8): 765-770

[34] Yakovlenko SI. Mechanism for the void formation in the bright spot of a fiber fuse. Laser Physics. 2006;**16**(3): 474-476

[35] Kawabe T, Kawai Y, Takayama K. Interaction between high frequency oscillations and low frequency oscillations in beam-plasma system. Journal of the Physical Society of Japan. 1967;**23**:1430-1431

[36] Abrams RH Jr, Yadlowsky EJ, Lashinsky H. Periodic pulling and turbulence in a bounded plasma. Physical Review Letters. 1969;**22**(7): 275-278

[37] Stix TH. Finite-amplitude collisional drift waves. Physics of Fluids. 1969; **12**(3):627-639

[38] Keen BE, Fletcher WHW. Suppression and enhancement of an ion-sound instability by nonlinear resonance effects in a plasma. Physical Review Letters. 1969;**23**(14):760-763

[39] Keen BE, Fletcher WHW. Suppression of a plasma instability by the method of asynchronous quenching. Physical Review Letters. 1970;**24**(4): 130-134

[40] Keen BE. Interpretation of experiments on feedback control of a "drift-type" instability. Physical Review Letters. 1970;**24**(6):259-262

[41] Shut'ko AV. Finite amplitude ion acoustic waves in an unstable plasma. Soviet Physics-JETP. 1970;**30**(2): 248-251

[42] Nakamura Y. Suppression of two-stream instability by beam modulation. Journal of the Physical Society of Japan. 1979;**28**(5):1315-1321

[43] Keen BE, Fletcher WHW. Remote feedback stabilization of the ion-sound instability by a modulated source at the electron-cyclotron resonance frequency. Physical Review Letters. 1970;**25**(6): 350-353

[44] Keen BE, Fletcher WHW. Measurement of growth rate, non-linear saturation coefficients, and mode-mode coupling coefficients of a 'Van der Pol' plasma instability. Journal of Physics D. 1970;**3**:1868-1885

[45] Nakamura Y. Suppression and excitation of electron oscillation in a beam-plasma system. Journal of the Physical Society of Japan. 1971;**31**(1): 273-279

[46] Keen BE, Fletcher WHW. Nonlinear plasma instability effects for subharmonic and harmonic forcing oscillations. Journal of Physics A. 1972;**5**: 152-165

[47] Boswell RW, Christiansen PJ, Salter CR. Nonlinear effects in an r.f. plasma. Physics Letters A. 1972;**38**(2):67-68

[48] Tavzes R, Cercek M. Frequency entrainment of a drift instability by nonlinear effects in a plasma. Physics Letters A. 1973;**43**(2):99-100

[49] DeNeef P, Lashinsky H. Van der Pol model for unstable wave on a beam-plasma system. Physical Review Letters. 1973;**31**(17):1039-1041

[50] Keen BE, Fletcher WHW. The ion-sound instability and its associated multi-mode phenomena. Journal of Physics D. 1973;**6**:1684-1698

[51] Michelsen P, Pecseli HL, Rasmussen JJ, Schrittwieser R. The current-driven, ion-acoustic instability in a collisionless plasma. Plasma Physics. 1979;**21**:61-73

[52] Ohno N, Tanaka M, Komori A, Kawai Y. Chaotic behavior of current-carrying plasmas in external periodic oscillations. Journal of the Physical Society of Japan. 1989;**58**(1):28-31

[53] Buragohain A, Chutia J, Nakamura Y. Mode suppression and period-doubling cascade in a double-plasma device. Physics Letters A. 1992;**163**: 425-428

[54] Gyergyek T, Cercek M, Jelic N, Stanojevic M. Mode suppression of a two-dimensional potential relaxation instability in a weakly magnetized discharge plasma. Physics Letters A. 1993;**176**(1):54-60

[55] Klinger T, Piel A, Seddighi F, Wilke C. Van der Pol dynamics of ionization plasma. Physics Letters A. 1993;**182**(2–3):312-318

[56] van der Pol B. The nonlinear theory of electric oscillations. Proc. Inst. Radio Engineering. 1934;**22**(9): 1051-1086

[57] Shuto Y. Heat conduction modeling of fiber fuse in single-mode optical fibers. Journal of Photonics. 2014;**2014**: Article ID 645207-1–21

[58] Shuto Y, Yanagi S, Asakawa S, Kobayashi M, Nagase R. Fiber fuse phenomenon in step-index single-mode optical fibers. IEEE Journal of Quantum Electronics. 2004;**40**(8):1113-1121

[59] Bumarin ED, Yakovlenko SI. Temperature distribution in the bright spot of the optical discharge in an optical fiber. Laser Physics. 2006;**16**(8): 1235-1241

[60] Facão M, Rocha A, André P. Traveling solution of the fuse effect in optical fibers. IEEE/OSA Journal Lightwave Technology. 2011;**29**(1): 109-114

[61] Todoroki S. Threshold power reduction of fiber fuse propagation through a white tight-buffered single-mode optical fiber. IEICE Electronics Express. 2011;**8**(23): 1978-1982

[62] Boyd RW. Nonlinear Optics. Chap. 1. New York: Academic Press, Inc.; 1992

[63] Okuyama A. Material Dynamics (Augmented Version). Chap. 5. Tokyo: Corona-sha; 1958

[64] Yokobori T. An Interdisciplinary Approach to Fracture and Strength of Solids. Chap. 1. Amsterdam: Gordon & Breach Publishers; 1968

[65] Schick HL. A thermodynamic analysis of the high-temperature vaporization properties of silica. Chemical Reviews. 1960;**60**:331-362

[66] Shuto Y. Evaluation of high-temperature absorption coefficients of ionized gas plasmas on optical fibers. IEEE Photonics Technology Letters. 2010;**22**(3):134-136

[67] JANAF. Thermochemical Tables. 2nd ed. U.S. Department of Commerce and National Bureau of Standards; 1971

[68] Saha MN. Ionization in solar chromosphere. Philosophical Magazine. 1920;**40**(238):472-488

[69] Kittel C. Introduction to Solid State Physics. Chap. 3. 7th ed. New York: John Wiley & Sons, Inc.; 1996

[70] Castellan GW. Physical Chemistry. Chap. 33. 3rd ed. Boston: Addison-Wesley Publishing Co., Inc.; 1983

[71] Chemical Data Book. Chap. 9. 3th ed. Tokyo: Chemical Society of Japan; 1984

[72] Shuto Y. End face damage and fiber fuse phenomena in single-mode fiber-optic connectors. Journal of Photonics. 2016;**2016**: Article ID 2781392-1-11

[73] Moore WJ. Physical Chemistry. Chap. 4. 4th ed. Princeton: Prentice Hall, Inc; 1972

[74] Brennen CE. Cavitation and Bubble Dynamics. Chap. 1. New York: Cambridge University Press; 2014

[75] Schottky W. Diffusionstheorie der positiven Säule. Physikalishce Zeitschrift. 1924;**25**:635-640

[76] Allis WP, Rose DJ. The transition from free to ambipolar diffusion. Physics Review. 1954;**93**(1):84-93

[77] Phelps AV. The diffusion of charged particles in collisional plasmas: Free and ambipolar diffusion at low and moderate pressures. Journal of Research of the National Institute of Standards and Technology. 1990;**95**(4):407-431

[78] Ohtori S, Sekiguchi T, Kawano T. Theory of Discharged Gases. Chap. 2. Tokyo: Ohmsha Inc.; 1969

[79] Israelachvili JN. Intermolecular and Surface Forces. Chap. 12. 2nd ed. London: Academic Press, Ltd.; 1992

[80] Kawakubo T, Kabashima S. Stochastic processes in self-excited oscillation. Journal of the Physical Society of Japan. 1974;**37**(5):1199-1203

Section 4

Optical Fiber Communication

Chapter 5

Delta-Sigma Digitization and Optical Coherent Transmission of DOCSIS 3.1 Signals in Hybrid Fiber Coax Networks

Jing Wang, Zhensheng Jia, Luis Alberto Campos and Curtis Knittle

Abstract

We first demonstrate delta-sigma digitization and coherent transmission of data over cable system interface specification (DOCSIS) 3.1 signals in a hybrid fiber coax (HFC) network. Twenty 192-MHz DOCSIS 3.1 channels with modulation up to 16384QAM are digitized by a low-pass cascade resonator feedback (CRFB) delta-sigma analog-to-digital converter (ADC) and transmitted over 80 km fiber using coherent single-λ 128-Gb/s dual-polarization (DP)-QPSK and 256-Gb/s DP-16QAM optical links. Both one-bit and two-bit delta-sigma digitization are implemented and supported by the QPSK and 16QAM coherent transmission systems, respectively. To facilitate its practical application in access networks, the coherent system is built using a low-cost narrowband optical modulator and RF amplifiers. Modulation error ratio (MER) larger than 50 dB is successfully demonstrated for all 20 DOCSIS 3.1 channels, and high order modulation up to 16384QAM is delivered over fiber for the first time in HFC networks. The raw DOCSIS data capacity is 54 Gb/s with net user information ~45 Gb/s. Moreover, the bit error ratio (BER) tolerance is evaluated by measuring the MER performance as BER increases. Negligible MER degradation is observed for BER up to 1.5×10^{-6} and 1.7×10^{-4}, for one-bit and two-bit digitization, respectively.

Keywords: access network, delta-sigma ADC, digitization, DOCSIS 3.1, hybrid fiber coax, OFDM

1. Introduction

Video-intensive services, such as virtual reality and immersive applications are driving the growth of data traffic at user premises in an explosive way, making access networks become a bottleneck of user quality of experience. Various optical and wireless access technologies have been investigated, including passive optical networks (PON) [1–3], cloud-radio access networks (C-RAN) [4–7], and hybrid fiber coax (HFC) networks [8, 9]. In the United States, there are more than 50 million subscribers using cable services for broadband access, which is 40% more than digital subscriber line (DSL) and fiber-to-the-home (FTTH) users [10]. Given the emergence of data over cable service interface specification (DOCSIS)

3.1, it is expected that HFC networks will continue to dominate the broadband access market in the United States, delivering fastest access speed to the broadest population.

As a fifth-generation broadband access technology, DOCSIS 3.1 specifications are being commercialized at a historically rapid pace to support ultra-high-resolution videos (4 K/8 K), mobile backhaul/fronthaul (MBH/MFH), and other emerging applications enabled by virtual reality and internet of things [10–14]. DOCSIS 3.1 specifications involve enhancement in both physical and MAC layers, which transform the physical layer signal from single-carrier QAM (SC-QAM) to orthogonal frequency division multiplexing (OFDM), for increased data rate, improved spectral efficiency, and flexible resource allocation. It provides up to 10 Gb/s downstream and 1.8 Gb/s upstream capacities to each subscriber [15, 16]. With subcarrier spacing of 25 or 50 kHz, DOCSIS 3.1 specifications support downstream channel bandwidths 24–192 MHz, and upstream channel bandwidths 6.4–96 MHz [17–19]. Moreover, higher order modulations up to 4096QAM were adopted with optional support of 8192 and 16384QAM [10, 15]. Similar to the LTE carrier aggregation in MFH networks [20, 21], DOCSIS 3.1 specifications support channel bonding to designate more than one DOCSIS channels to a single user.

The continuous envelope and high peak-to-average power ratio (PAPR) of OFDM signals, on the other hand, make them vulnerable to noise and nonlinear impairments in analog HFC networks [22–24]. Combined with demanding carrier-to-noise ratio (CNR) requirements of high order modulations (>4096QAM), it is difficult to support DOCSIS 3.1 signals by legacy analog fiber links [15]. In this paper, we for the first time demonstrate the digitization of DOCSIS 3.1 signals to enable the upgrade of fiber distribution networks from analog to digital, so mature digital fiber technologies, e.g., intensity modulation/direct detection (IM/DD) and coherent optical transmission, can be exploited.

To enable digital transmission of DOCSIS 3.1 signals, a digitization interface, i.e., analog-to-digital converter (ADC), is needed in the hub to digitize the analog signals into bits, and a digital-to-analog converter (DAC) is needed in the fiber node to retrieve the analog waveforms from digital bits for the following transmission over coaxial cable plant. Different from conventional Nyquist AD/DA that uses Nyquist sampling rates, such as common public radio interface (CPRI) in MFH networks [25], which has quantization noise evenly distributed in the frequency domain and needs many quantization bits, delta-sigma ADC features high sampling rate but only a few (one or two) quantization bits, and most importantly, it utilizes a noise shaping technique to push the quantization noise out of the signal band, so that signal and noise are separated in the frequency domain, and the in-band CNR of digitized signals can be optimized [26–29]. Moreover, a simplified DAC design based on low-cost passive filters can be used in the fiber node, which filters out the desired signals, eliminates the out-of-band noise, and at the same time, retrieves the analog waveforms. In the hub, a high-speed delta-sigma ADC is shared by multiple fiber nodes; whereas in each fiber node, only a low-cost passive filter is needed to filter out the desired signal and convert it to the analog waveform. Since there are more fiber nodes than hubs, especially given the fact that fiber node number is continuing to grow due to node segmentation and fiber deep migration, replacing Nyquist DAC with a low-cost passive filter can significantly reduce the cost and complexity of fiber nodes.

Delta-sigma digitization has found wide applications in power amplifiers [30–32], RF transmitters [33–37] and receivers [38–42], visible light communications [43, 44], radio-over-fiber (RoF) [45–47], and MFH networks [48–50]. In Ref. [48-50], we first demonstrated delta-sigma digitization as a new digitization interface in MHF networks to replace CPRI, and 32 LTE carrier aggregation was demonstrated within a single-λ 10 Gb/s PON system to support 3GPP release 13. We then extended delta-sigma digitization to DOCSIS signals for HFC networks [51]. This paper is an extended version of our previous work [51] with substantial details and new results.

In this paper, we for the first time demonstrate the delta-sigma digitization of twenty 192-MHz DOCSIS 3.1 channels with 16384QAM modulation, based on a low-pass cascade resonator feedback (CRFB) delta-sigma ADC. We transmit the digitized bits over 80 km fiber by a low-cost single-λ 128-Gb/s dual-polarization (DP)-QPSK or 256-Gb/s DP-16QAM coherent fiber link. Both one-bit and two-bit delta-sigma digitization are realized and supported by the coherent QPSK/16QAM links, respectively. To facilitate its application in access networks, the coherent fiber link is built using low-cost narrowband RF amplifiers and optical modulator. More than 50 dB modulation error ratio (MER) is achieved for all 20 DOCSIS 3.1 channels, and high order modulation up to 16384QAM is demonstrated and delivered over fiber for the first time in HFC networks. The raw DOCSIS data rate is 54 Gb/s with net user information ~45 Gb/s. The bit error ratio (BER) tolerance of delta-sigma digitization is also evaluated and negligible MER performance degradation is observed for BER values up to 1.5×10^{-6} and 1.7×10^{-4}, for one-bit and two-bit digitization, respectively.

This chapter is organized as follows. Section 2 discusses the operation principles of delta-sigma digitization. Section 3 presents the experimental setup. Section 4 shows the design of delta-sigma ADC. The experimental results are shown in Section 5. Section 5 concludes the paper.

2. Operation principles

The architecture of a HFC network is shown in **Figure 1**. Due to their similarity, a C-RAN architecture is also presented for comparison. In **Figure 1(b)**, the network segment from service gateway (S-GW) or mobile management entity (MME) to

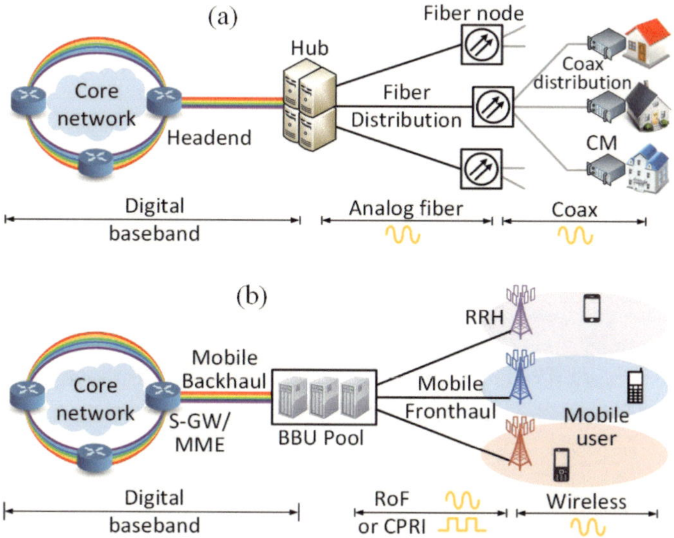

Figure 1.
Architecture of HFC network and C-RAN: (a) HFC network and (b) C-RAN.

baseband unit (BBU) is defined as mobile backhaul (MBH), which transmits the control and payload bits of LTE signals in digital baseband. The digital bits are received by BBUs, which synthesizes OFDM modulation and generates analog waveform of LTE signals. The network segment from BBU to remote radio heads (RRH) is defined as mobile fronthaul (MFH), where the LTE signals are transmitted over fiber in either analog waveform using RoF technology or digital waveform using CPRI digitization interface. The last segment of C-RAN involves the wireless transmission from the RRHs to mobile users, where LTE signals are transmitted in their analog waveforms via air interface.

Similarly, HFC networks in **Figure 1(a)** can also be divided into three segments, i.e., the core network segment from headend to hub, the fiber distribution network from hub to fiber nodes, and the coaxial cable plant from fiber nodes to cable modems (CMs). Similar to MBH, the segment from headend to hub transmits net bit information; similar to MFH, the fiber distribution segment from hub to fiber node is supported by either analog or digital fiber technologies, e.g., C-RAN uses RoF technology to deliver analog mobile signals; HFC uses analog fiber links to deliver analog DOCSIS/video signals; C-RAN uses CPRI as a Nyquist digitization interface; HFC has a similar interface called baseband digital forward or return (BDF/BDR). The last segment from fiber node to CMs is also similar to the wireless segment of C-RAN, where both DOCSIS and LTE signals are transmitted in their analog waveform over coaxial cable or air interface, respectively. Different analog or digital implementations of the fiber distribution segment are shown in **Figure 2**.

2.1 Analog fiber

Figure 2(a) shows the architecture of an analog fiber link. DOCSIS and video signals are aggregated in the hub and delivered to the fiber node in analog waveforms. Then at the fiber node, the received analog signals are delivered to CMs via cable distribution networks. An analog fiber link features simple, low-cost implementation, and high spectral efficiency, but imposes high linearity requirements on channel response. Since it does not perform any data reformation, an analog fiber link is a waveform/service agnostic pipe, and can be used for various services, e.g., DOCSIS, MPEG, and analog TV. On the other hand, it suffers from noise and nonlinear impairments, limited signal-to-noise ratio (SNR), short fiber distance, and small number of WDM wavelengths. It also requires complex RF amplifiers and bi-annual calibration of fiber nodes.

2.2 BDF/BDR

Upgrading fiber distribution networks from analog to digital offers the opportunity to leverage the existing mature digital access technologies. Due to the wide

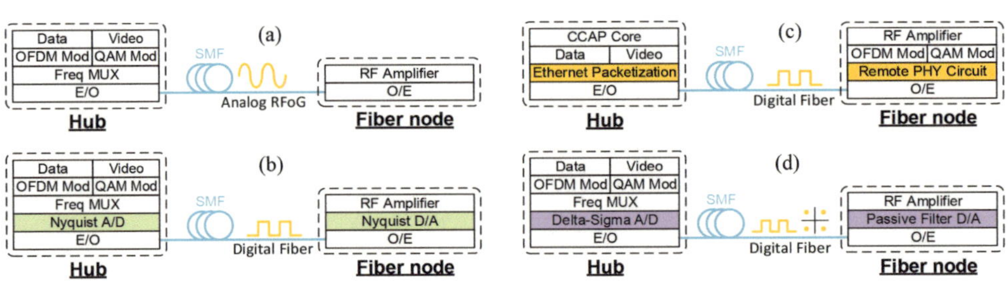

Figure 2.
Different analog/digital technologies for fiber distribution network: (a) analog fiber; (b) BDF/BDR; (c) remote PHY; (d) delta-sigma digitization.

deployment of Ethernet, digital fiber link features low cost, high capacity, long fiber distance, and easy setup/maintenance, as compared with its analog counterpart. Moreover, the contribution of optical noise and nonlinear impairments can be isolated from the received signal quality, so large SNR and high order modulation can be achieved. Since it can support many (>80) WDM wavelengths, digital fiber link facilitates the migration to node split and fiber deep. **Figure 2(b-d)** shows three different digital fiber technologies.

Figure 2(b) shows a digital fiber link based on BDF/BDR architecture, where a Nyquist ADC is inserted in the hub with 2.5 oversampling ratio and 12 quantization bits. At fiber node, a Nyquist DAC retrieves the analog signals, and feeds them to the coaxial cable plant. Same as CPRI, BDF/BDR provides a simple, low-cost, and service-transparent digitization interface, but with low spectral efficiency. Similar to CPRI, the digitized bits of BDF/BDR are not encapsulated into Ethernet packets but framed by time-division-multiplexing (TDM), so it always runs at full data rate even without any real payload. This makes traffic engineering based on statistical multiplexing impossible. In today's HFC networks, there is only upstream BDR deployed, but no downstream BDF, and the BDR specifications are vendor proprietary and not interoperable.

2.3 Remote PHY

Digital fiber link based on remote PHY architecture is shown in **Figure 2(c)**, where the PHY chips for OFDM/QAM modulation and demodulation are moved to fiber node, and an integrated converged cable access platform (CCAP) is separated into the CCAP core in hub and the remote PHY device (RPD) in fiber node [52–56]. In the downstream, payload and control bits are packetized into Ethernet packets and transmitted from hub to fiber node, where the OFDM/QAM modulators synthesize analog DOCSIS/MPEG signals for cable distribution. In the upstream, OFDM/QAM demodulators interpret the received analog signals into baseband bits and transmit back to hub in Ethernet packets. Compared with analog fiber link in **Figure 2(a)**, the RF interface in hub is replaced by an Ethernet interface, and in fiber node, there is an Ethernet interface connecting to the digital fiber and a RF interface connecting to the coaxial cable plant. With the help of Ethernet packetization, remote PHY architecture can exploit Ethernet access technologies, such as Ethernet PON (EPON), gigabit PON (GPON), and Metro Ethernet [52–54], and enable statistical multiplexing for traffic engineering. Compared with other digital solutions, remote PHY features smaller traffic load in the fiber, but with the penalty of increased complexity and cost of fiber nodes. Due to the modulation/demodulation at RPD, the fiber link in remote PHY architecture is no longer a service-transparent pipe, although it maintains the least amount of hardware exported to RPD, and preserves the compatibility with existing hubs in analog fiber links. It should be noted that the concepts of remote PHY/MAC are very similar to the function split of MFH networks [57–59], by moving partial physical and/or MAC layer functions from the centralized entity (hub/BBU) to a remote node (fiber node/RRH).

2.4 Delta-sigma digitization

Figure 2(d) shows the architecture of delta-sigma digitization. Compared with **Figure 2(b)**, Nyquist AD/DA in BDF/BDR are replaced by a delta-sigma ADC in hub and a passive filter in fiber node. Different from the Nyquist ADC with oversampling ratio of 2.5 and 12 quantization bits, delta-sigma ADC trades quantization bits for sampling rate, using high sampling rate but only a few (one or two) quantization bits. Its operation principle is shown in **Figure 3**. For reference, the operation principle of Nyquist ADC is also presented in **Figure 3(a)**. In this paper, we designed a delta-sigma ADC to digitize five DOCSIS 3.1 channels with channel bandwidth of 192 MHz and total frequency range from 258 to 1218 MHz (5 × 192 = 960 MHz). Due to the limited quantization bits, Nyquist sampling rate will lead to significant quantization noise (**Figure 3b**). Oversampling is utilized to extend Nyquist zone and spread the quantization noise over a wide frequency range, so the in-band noise is reduced (**Figure 3c**).

Furthermore, noise shaping technique is used to push the quantization noise out of the signal band, which acts as a high-pass filter (HPF) to the quantization noise and separates the signal and noise in the frequency domain, as shown in **Figure 3(d)**. It should be noted that during the delta-sigma digitization, the signal spectrum is kept intact; it is the out-of-band quantization noise added by the delta-sigma ADC that converts the signal waveform from analog to digital. Therefore, in **Figure 3(e)** at the receiver side, when the out-of-band quantization noise is eliminated, the signal waveform will automatically be converted back from digital to analog, i.e., a passive filter can not only filter out the desired signal channel, but also realize the digital-to-analog conversion. In HFC networks, a high-speed delta-sigma ADC is centralized in the hub and shared by many fiber nodes, and each fiber node only needs a low-cost passive filter to select the desired DOCSIS channels and at the same time convert them to the analog waveform. Given the fact that there are many more fiber nodes than hubs/headends, this design significantly reduces the cost and complexity of fiber nodes. It should be note that Nyquist ADC has evenly distributed quantization noise, whereas delta-sigma digitization has a shaped distribution of quantization noise, so the retrieved analog signal has an uneven noise floor.

In time domain, if we use an analog sinusoidal signal as an example, a Nyquist ADC samples the analog input with Nyquist rate and each sample is quantized individually (**Figure 3a**); whereas delta-sigma ADC samples the analog input at a much higher rate and the samples are digitized consecutively (**Figure 3c** and **d**), i.e., the current digitization bits not only depend on the current analog input, but also depend on previous input. One-bit delta-sigma digitization outputs a high data rate on-off keying (OOK) signal with the density of "1" bits being proportional to the amplitude of analog input. For maximum input, it outputs continuous "1"s; for minimum input, it outputs continuous "0"s. For intermediate inputs, the densities of "0"s and "1"s are almost equal. Two-bit digitization outputs a 4-level pulse-amplitude-modulation (PAM4) signal. Both one-bit and two-bit digitization are implemented in our experiment.

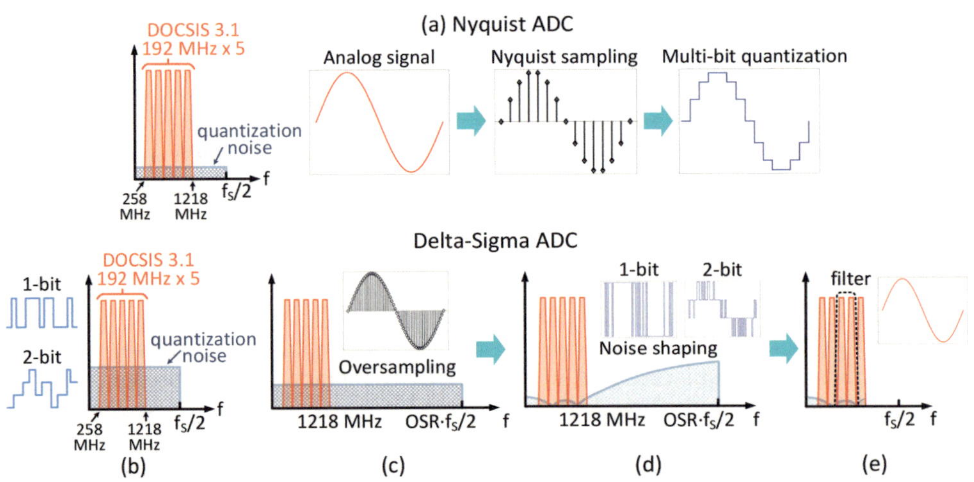

Figure 3.
Operation principle of Nyquist ADC and delta-sigma ADC: (a) Nyquist ADC and (b-e) delta-sigma ADC.

	Analog	Digital		
Interface	Linear optics	BDF/BDR	Remote PHY	Delta-sigma digitization
Operation principles	Analog RF over fiber	Nyquist AD/DA in hub and fiber node	Move PHY circuits of OFDM/QAM to fiber node Transmit net bit information over fiber	Delta-sigma ADC in hub Passive filter in fiber node as DAC
AD/DA	N/A	2.5 oversampling ratio 12 quantization bits	N/A	High sampling rate 1–2 quantization bits
Pros	Simple High spectral efficiency Waveform agnostic	High capacity, Large SNR, High order modulation, Long distance, Scalability, Many WDM wavelengths, facilitate node split and fiber deep migration		
		No modification of bits Waveform agnostic Simple, low-cost AD/DA	Ethernet packet encapsulation Statistical multiplexing Reduced traffic load	No modification of bits Waveform agnostic Low-cost DA based on passive filters
Cons	Short distance Small capacity Limited by noise/nonlinearity Few WDM wavelengths	Low spectral efficiency No Ethernet encapsulation Always run at full data rate Vendor proprietary	Modification of bits Not service transparent Increased complexity/cost of fiber nodes	High cost delta-sigma ADC

Table 1.
Comparison of analog/digital fiber technologies in HFC networks.

Table 1 compares different analog/digital fiber technologies in HFC networks. As a waveform-agnostic interface, delta-sigma ADC works with not only OFDM signals but also 5G multicarrier waveforms, such as filter-band multicarrier signals, as we reported in [50]. Since analog fiber, BDF/BDR, and delta-sigma digitization do not modify the bit information, they are service-agnostic and can carry various and a combination of services, even though these services evolve in the future. Remote PHY, on the other hand, is not service-transparent. Its RPD in the fiber node is only designed for one specific service.

3. Experimental setup

Figure 4 shows the experimental setup. OFDM parameters of DOCSIS 3.1 specifications are listed in **Table 2**. There are two sets of FFT sizes and subcarrier spacing. In this experiment, we use the FFT size of 4096 and subcarrier spacing of 50 kHz.

In a 1.2 GHz DOCSIS 3.1 implementation, the downstream signal contains five channels, each with 192-MHz bandwidth and occupying 960-MHz (258–1218 MHz) frequency band in total. In this experiment, these five downstream channels are digitized by delta-sigma ADCs with sampling rates of 16, 20, 24, 28, and 32 GSa/s. Both one-bit and two-bit digitization are carried out, and the five channels are digitized to an OOK (one-bit) or PAM4 (two-bit) signal with baud rate of 16–32 Gbaud. In a dual-polarization (DP) coherent fiber link, each polarization has in-phase (I) and quadrature (Q) components, and each component can carry one OOK/PAM4 signal, so there are four data streams in total carrying 20 digitized DOCSIS 3.1 channels, i.e., a DP-QPSK/16QAM link can carry 20 digitized DOCSIS 3.1 channels with one-bit or two-bit digitization, respectively. Delivering 20 DOCSIS channels over a single wavelength quadruples the capacity of current HFC networks and enables a 4x1 fiber node split. In the following sections, only the results of 32 Gbaud are discussed in detail due to the limited space.

In **Figure 4**, the two arms of IQ Mach-Zehnder modulator (MZM) are driven by two independent OOK/PAM4 signals to synthesize a QPSK/16QAM signal, and

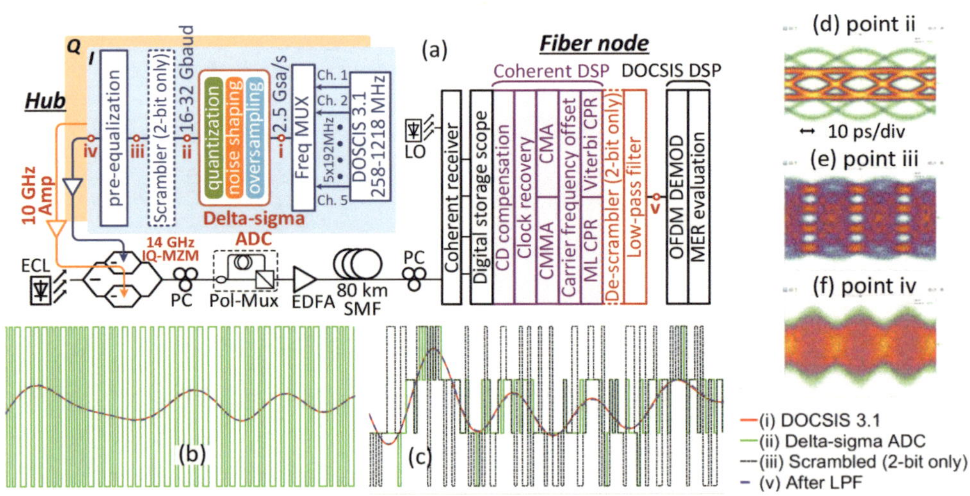

Figure 4.
(a) Experimental setup; (b) waveforms for one-bit delta-sigma digitization; (c) waveforms for two-bit delta-sigma digitization; (d) eye diagram of 32 Gbaud PAM4 signal after delta-sigma ADC (point ii); (e) PAM4 eye diagram after scrambling (point iii); (f) PAM4 eye diagram after pre-equalization (point iv).

Sampling rate	FFT size	Subcarrier spacing	Active subcarriers	Active BW	Data subcarriers	Data BW	Guard band
204.8 MSa/s	4096	50 kHz	3840	192 MHz	3800	190 MHz	1 MHz
	8192	25 kHz	7680		7600		

In experiments, we use FFT size of 4096 and subcarrier spacing of 50 kHz.

Table 2.
OFDM parameters of DOCSIS 3.1 signals.

after the polarization multiplexer, a DP-QPSK/16QAM signal carries four streams of OOK/PAM4 with totally 20 digitized DOCSIS channels.

It is worth noting that although there are several reports of high speed delta-sigma ADC with sampling rate up to 8.6 GSa/s [35–37], there is no commercially available delta-sigma ADC that runs faster than 10 GSa/s. For a proof-of-concept experiment in this paper, the delta-sigma digitization is realized offline using MATLAB, and the digitized bits are loaded into a Keysight arbitrary waveform generator (AWG) M8196A, and then transmitted over a 80-km coherent fiber link. In real implementations, to alleviate the speed limit, several low speed delta-sigma ADCs can be used in parallel, each digitizing only one DOCSIS channel, rather than using a high-speed ADC to digitize all five channels together. The output bits from parallel low speed ADCs can be interleaved in the time domain by TDM technology, so the sampling rate of each ADC is reduced, while keeping the overall capacity intact.

In **Figure 4(b)** and **(c)**, analog DOCSIS 3.1 signal at point i is plotted in red; after delta-sigma digitization, OOK/PAM4 signals at point ii are plot in green; retrieved analog signals after filters at point v are plotted in dashed blue lines. The initial (red) and retrieved (dashed blue) analog signals are fairly close to each other, indicating that the digitization introduces almost no impairment. In the green curve of **Figure 4(c)**, there are more ±1 symbols than ±3 symbols. This is because DOCSIS 3.1 is an OFDM signal with Gaussian distribution, and there are much more small samples than large ones. Therefore, the PAM4 signal after digitization also has unequal distribution. More than 80% symbols are ±1 s, and only less than 20% are ±3 s. This also makes the 16QAM signal has unequal distribution on the constellation. Most symbols are at (±1 ± j), and only a few at (±3 ± 3j). Unequally distributed constellation introduces challenges to the digital signal processing (DSP) of the coherent receiver, especially for constant multiple modulus algorithm (CMMA). To equalize the symbol distribution, a scrambler is inserted in the transmitter (only for two-bit digitization). Eye diagrams before and after the scrambler are shown in **Figure 4(d)** and **(e)**. In **Figure 4(d)**, there are much more ±1 s than ±3 s; in **Figure 4(e)**, the amount of ±1 s and ± 3 s are equalized.

Since delta-sigma digitization is designed to be utilized in access networks, such as HFC and C-RAN, a low-cost coherent system was built based on narrowband devices, e.g., 10 GHz RF drivers (Picosecond Pulse Labs 5822B) and 14 GHz IQ-MZM (Covega LN86S-FC). For 32 Gbaud QPSK/16QAM, these narrowband devices significantly impair the transmission performance, and frequency domain pre-equalization was used to compensate the bandwidth limitation. **Figure 4(f)** shows the 32 Gbaud PAM4 eye diagram after pre-equalization. The eye is closed due to the boosted high frequency components.

After 80-km single mode fiber, the DP-QPSK/16QAM signal is received at the fiber node. In experiments, a Keysight N4391A optical modulation analyzer is used as a polarization diversity receiver. Four received signals (two polarizations, each polarization has in-phase and quadrature components) are captured by a real-time

QAM	16	64	128	256	512	1024	2048	4096	8192*	16384*
CNR (dB)	15	21	24	27	30.5	34	37 (37.5)+	41 (41.5)+	44 (44.5)+	48 (48.5)+

**For DOCSIS 3.1 downstream, 8192QAM and 16384QAM are optional, and their CNRs are not specified yet. Here we use 44(44.5) and 48(48.5) dB as temporary criteria.*
+CNR values in parentheses with 0.5-dB increment are for channels above 1 GHz.

Table 3.
Carrier-to-noise ratio (CNR) requirement of DOCSIS 3.1 specifications.

digital storage oscilloscope Keysight DSAX92004A for offline DSP. We use standard coherent DSP algorithms, including Gram-Schmidt orthogonalization [60], chromatic dispersion (CD) compensation [61, 62], polarization de-multiplexing [63, 64], carrier frequency offset (CFO) recovery [65, 66], and carrier phase recovery (CPR) [67, 68]. For polarization de-multiplexing, QPSK uses constant modulus algorithm (CMA), 16QAM uses constant multiple modulus algorithm (CMMA). For CPR, QPSK uses Viterbi-Viterbi algorithm, 16QAM uses the maximum likelihood (ML) phase recovery algorithm. After coherent DSP, a de-scrambler is applied to the PAM4 signal to restore its initial symbol distribution, and five DOCSIS channels are filtered out by a digital filter to retrieve their analog waveforms.

To evaluate the performance of delta-sigma digitization and coherent transmission, we use carrier-to-noise ratio (CNR) of received DOCSIS channels as a measurement. Required CNR for different modulations in DOCSIS 3.1 specifications are listed in **Table 3** [15]. Higher order modulations need higher CNR, and there is 0.5 dB increment for the fifth channel above 1 GHz (1026–1218 MHz). The maximum mandatory modulation in DOCSIS 3.1 specification is 4096QAM. 8192 and 16384QAM are optional, and their CNR requirements are not specified yet. Here we use 44 and 48 dB based on extrapolation. In experiments, CNR is evaluated in terms of modulation error ratio (MER).

4. Experimental design

In this section, we first present the design of delta-sigma ADC and discuss the experimental design, then demonstrate the results of one-bit and two-bit digitization, respectively. Finally, the performance tolerance against the bit error ratio (BER) of the coherent fiber link is also evaluated.

4.1 Delta-sigma ADC design

The design of a fourth-order delta-sigma ADC based on cascaded resonator feedback (CRFB) structure is shown in **Figure 5**. The Z-domain block diagram is shown in **Figure 5(a)**, which consists of an output quantizer, a feedback DAC, and the rest parts can be considered as a filter to the quantization noise. The transfer function of this noise filter is described by a noise transfer function (NTF), which determines the frequency distribution of quantization noise. In **Figure 5(a)**, there are four integrators, and every two of them, $1/(z-1)$ and $z/(z-1)$, are cascaded together to form a resonator (purple and green). The number of integrators equals to the order number of the NTF. A fourth-order NTF also has two conjugate pairs (four in total) of zeroes and poles, shown in **Figure 5(b)**. The frequency response of the NTF is shown in **Figure 5(c)**.

Our design is a low-pass delta-sigma ADC where the signal is located at the low frequency end, and quantization noise is at the high frequency end. So in **Figure 5(c)**, the NTF is a high-pass filter, which pushes the quantization noise to

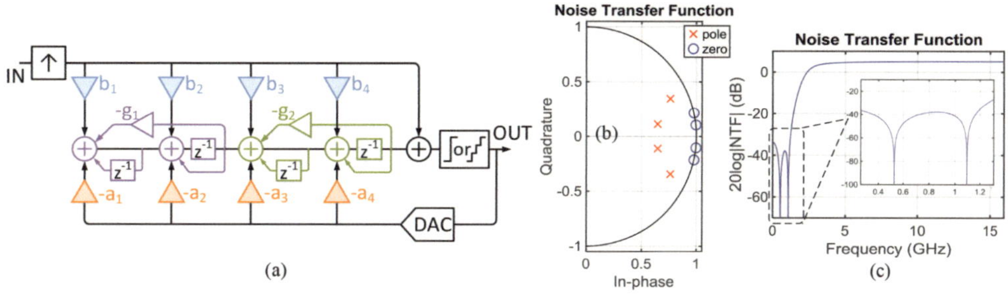

Figure 5.
Implementation of 32 GSa/s delta-sigma ADC. (a) Z-domain block diagram of a fourth-order cascade resonator feedback (CRFB) structure. (b) Zeroes and poles of the noise transfer function (NTF). (c) Frequency response of the NTF.

the high frequency end and separates it from the signal. In the inset of **Figure 5(c)**, there are two notches in the stopband of the NTF, each corresponding to one pair of zeroes in **Figure 5(b)**. At the zeroes of NTF, quantization noise is minimized and signals at these frequency points have a maximized CNR. It should be noted that the only difference of one-bit and two-bit digitization is the quantizer at the output and the feedback DAC. Their NTFs are identical. The number of output levels is determined by the number of quantization bits. A log2(N)-bit quantizer outputs N levels, so one-bit quantizer outputs an OOK signal, and two-bit quantizer outputs a PAM4 signals. More details of delta-sigma ADC design can be founded in Ref [28, 29].

4.2 Experimental cases

To evaluate the performance of delta-sigma digitization, 10 experimental cases are designed, shown in **Table 4**. Five DOCSIS 3.1 channels are digitized by delta-sigma ADCs with sampling rates of 16, 20, 24, 28, and 32 GSa/s. Both one-bit (Case I-V) and two-bit (Case VI-X) digitization are carried out, and the fiver DCOSIS channels are digitized to a 16–32 Gbaud OOK (one-bit) or PAM4 (two-bit) signal, respectively. The signal baud rate after digitization is equal to the sampling rate of ADC. In a dual-polarization coherent fiber link, each polarization has I and Q components, and each component carries one OOK/PAM4 signal, so there are four data streams in total carrying 20 digitized channels. Due to the symmetry, only 5 out of 20 DOCSIS 3.1 channels are listed in **Table 4**.

Unlike Nyquist ADC, whose quantization noise is evenly distributed in the Nyquist zone, delta-sigma ADC has uneven noise floor due to the noise shaping technique. In experiments, different modulations are assigned to different channels according to their CNRs, e.g., in Case V, only Ch. 2 has sufficient CNR to support 16384QAM, Ch. 4 can only support 4096QAM, and the rest three can carry 8192QAM. In general, higher sampling rate leads to wider Nyquist zone and smaller in-band quantization noise, so higher modulation can be supported. Two-bit digitization always has smaller quantization noise thanks to the additional bit. Therefore, in Case IX and X, all five channels have sufficient CNR to carry 16384QAM.

4.3 Comparison with Nyquist ADC

Spectral efficiency is an important figure of merit for digitization interfaces, and it is insightful to make a comparison of two digitization interfaces in terms of spectral efficiencies. Since DOCSIS 3.1 channels can support various modulations from 16QAM up to 4096QAM, the net data capacity per channel may vary dramatically;

Case	Sampling rate (GSa/s)	Quantization	Waveform after ADC	Coherent transmission	Modulation formats of five DOCSIS channels					BER
I	16	One-bit	OOK	QPSK	128	1024	128	128	128	Error free
II	20				256	2048	256	256	256	
III	24				1024	8192	1024	1024	1024	
IV	28				2048	16,384	2048	2048	2048	
V	32				8192	16,384	8192	4096	8192	
VI	16	Two-bit	PAM4	16QAM	512	4096	512	512	512	
VII	20				2048	16,384	2048	2048	2048	
VIII	24				4096	16,384	4096	4096	4096	1.2×10^{-5}
IX	28				16,384	16,384	16,384	16,384	16,384	2.8×10^{-5}
X	32				16,384	16,384	16,384	16,384	16,384	1.1×10^{-4}

Table 4.
Experimental design of one-bit and two-bit delta-sigma digitization.

but the frequency band of five downstream DOCSIS 3.1 channels are always the same, from 258 to 1218 MHz. Therefore, to make a fair comparison, we can use the number of bits needed to digitize five DOCSIS channels as a reference.

In **Table 4**, for one-bit delta-sigma ADC, depending on the sampling rate, 16–32 Gb/s fiber link capacity is needed to support five digitized channels. For Case V (32 GSa/s), all five channels can support at least 4096QAM, within which at least three of them can support 8192QAM and one can carry up to 16384QAM. For two-bit ADC, the required bit number doubles. On the other hand, consider the Nyquist ADC defined in BDR/BDF with 2.5 oversampling ratio and 12 quantization bits, it requires 1218 × 2.5 MSa/s × 12 b/Sa = 36.54 Gb/s to digitize five DOCSIS channels.

Compared with Nyquist ADC, one-bit delta-sigma ADC can save at least 12.4% (Case V) fiber link capacity. If CNR requirements can be relaxed, and lower sampling rate are allowed, it can save up to 56.2% (Case I) fiber link capacity. For two-bit delta-sigma ADC, although it provides much higher CNR, it generates more bits after digitization and consumes more fiber link capacity than Nyquist ADC. The main advantage of applying delta-sigma digitization in HFC networks is to simplify fiber node design by replacing the conventional DAC by a low-cost passive filter. Improvement of spectral efficiency is a side effect.

5. Experimental results

In this section, we first demonstrate the experimental results of DP-QPSK and DP-16QAM coherent transmission, and then discuss the results of one-bit and two-bit digitization supported by QPSK/16QAM, respectively. Finally, we will investigate the BER tolerance of delta-sigma digitization.

5.1 Coherent transmission

Figure 6 shows the results of coherent transmission of DP-QPSK/16QAM, where error vector magnitude (EVM) is plotted as a function of baud rate. Due to the limited bandwidth of low-cost RF amplifiers and the optical IQ MZM, EVM increases with baud rate. **Figure 6(b)** shows the constellations at each baud rate. For DP-QPSK, error free transmission can be achieved for all baud rates. For DP-16QAM, error free transmission is only achievable for 16 and 20 Gbaud. BER values for 24, 28 and 32 Gbaud are labeled in **Figure 6(a)**. In the following discussion, we only present the results of 32 GSa/s delta-sigma digitization. Results of other sampling rates are similar and omitted here for brevity.

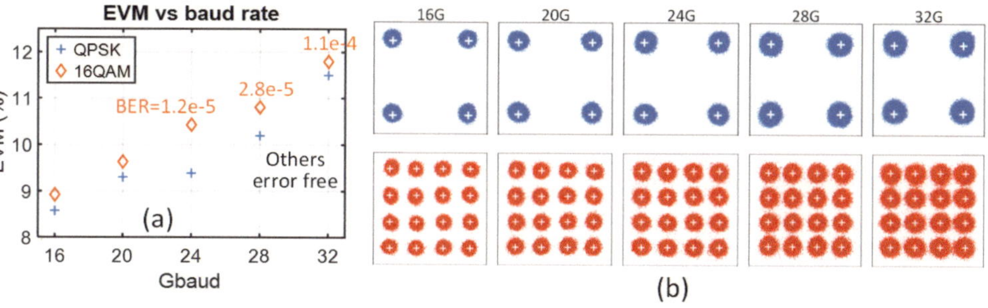

Figure 6.
Coherent transmission results of low-cost DP-QPSK and DP-16QAM systems, for one-bit and two-bit digitization, respectively. (a) Error vector magnitude (EVM) vs. baud rate. (b) Constellations of QPSK and 16QAM at each baud rate.

5.2 One-bit digitization

Figure 7 shows the experimental results of 32 GSa/s one-bit digitization (Case V). 20 DOCSIS 3.1 channels are digitized into four 32 Gb/s OOK signals, and then transmitted by a 32 Gbaud DP-QPSK link. The RF spectra before (blue) and after (red) delta-sigma digitization are shown in **Figure 7(a)**. Five 192-MHz DOCSIS 3.1 channels occupy a frequency range from 258 to 1218 MHz. It can be seen that after delta-sigma digitization, the signal spectrum is kept intact, and the quantization noise is pushed out of the signal band with in-band CNR larger than 40 dB. **Figure 7(b)** shows the MER performance of five channels, and each of them satisfies the requirements (dashed lines) of DOCSIS 3.1 specifications. Channel 2 has MER > 50 dB carrying 16384QAM; channel 4 has MER > 41 dB carrying 4096QAM; the remaining three have MER > 44 (44.5) dB carrying 8192QAM. This is the first time that modulation orders higher than 4096QAM have been demonstrated over fiber in HFC networks. Constellations of each modulation are shown in **Figure 7(c, d, e)**. For 8192QAM, the worst case, channel 3 with lowest MER of 44.5 dB, is shown.

5.3 Two-bit digitization

Figure 8 shows the experimental results of 32 GSa/s two-bit delta-sigma digitization (Case X). Five DOCSIS 3.1 channels are digitized into a 32 Gb/s PAM4 signal. The RF spectra before (blue) and after (red) digitization are shown in **Figure 8(a)**. Compared with **Figure 7(a)**, the in-band quantization noise is reduced thanks to the additional quantization bit, and CNR > 50 dB is achieved. In Case IX and X, with sampling rates of 28 and 32 GSa/s, all 20 DOCSIS 3.1 channels have enough CNR to support 16384QAM. The raw data rate of these 20 channels is 53.2 Gb/s, and the net user information is ~45 Gb/s. Since the minimum data capacity of a coherent fiber link to support 2016384QAM channels

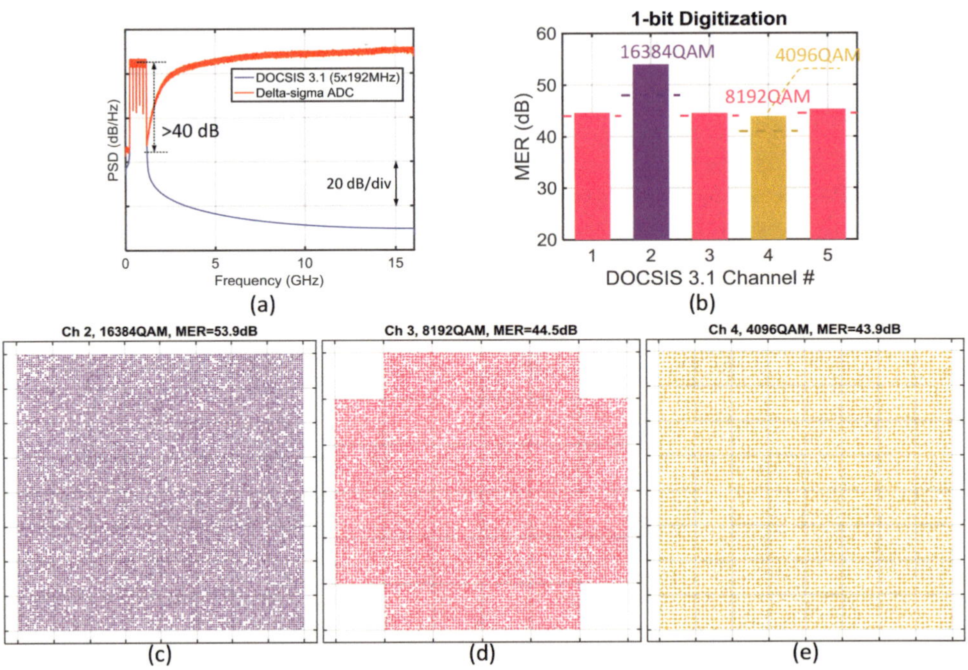

Figure 7.
Experimental results of 32 GSa/s one-bit delta-sigma digitization (Case V): (a) RF spectra; (b) MER of five DOCSIS 3.1 channels; (c) Channel 2 16384QAM; (d) channel 3 8192QAM; (e) channel 4 4096QAM.

is 28 × 4 × 2 = 224 Gb/s, the net information ratio of delta-sigma digitization is 45/224 = 20.1%, which is higher than BDF/BDR interface based on Nyquist digitization. **Figure 8(b)** shows the MER performance of 5 out of 20 DOCSIS channels, each with a MER larger than 52.5 dB, which is 4.5 dB higher than the required 48 dB for 16384QAM in DOCSIS 3.1 specifications. This indicates that two-bit digitization at 32 GSa/s has the potential to support an even higher modulation reaching up to 32768QAM. **Figure 8(c)** shows the worst-case constellation of 16384QAM, which corresponds to channel 4 with the lowest MER of 52.5 dB.

5.4 BER tolerance

In **Table 4** and **Figure 6**, error free transmission is achieved for all baud rate DP-QPSK (Case I–V); but for DP-16QAM, error free transmission is only achieved at 16 and 20 Gbaud (Case VI, VII). In this section, we design BER tolerance test to measure the MER performance degradation of delta-sigma digitization as a function of increasing BER, as shown by Case XI and XII in **Table 5**, where 4096QAM are assigned to all 20 channels since it is the highest modulation specified in DOCSIS 3.1 specifications (8192/16384QAM are optional).

Figure 9 shows the results of one-bit delta-sigma digitization (Case XI). The experimentally measured EVM and BER of 32 Gbaud DP-QPSK at different received optical power are presented in **Figure 9(a)**, with constellations plotted in the insets. Due to the limited memory of our AWG, the minimum measurable BER is 1×10^{-6}, and error free transmission is achieved for received optical power larger than −10 dBm. **Figure 9(b)** shows the MER degradation as a function of increasing BER. Both simulation (solid lines) and experiments (dots) are carried out. In the simulation, various BER values are emulated by flipping a certain number of bits after delta-sigma digitization; whereas in the experiments, different BER values are obtained by reducing the received optical power. In experiments, it is impossible to achieve exact BERs by adjusting the optical power. In **Figure 9(b)**, only three values

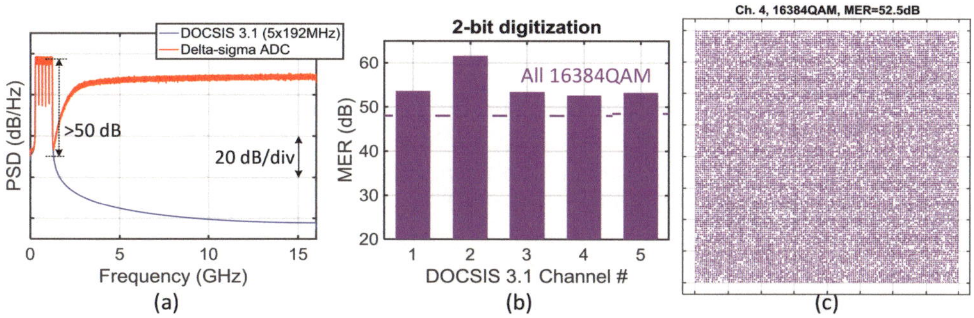

Figure 8.
Experimental results of 32 GSa/s two-bit delta-sigma digitization (Case X): (a) RF spectra; (b) MER performance of five DOCSIS 3.1 channels; (c) channel 4 16384QAM (worst of the five channels).

Case	Sampling rate (GSa/s)	Quantization	Waveform after ADC	Coherent transmission	Modulation formats of five DOCSIS channels				
XI	32	One-bit	OOK	QPSK	4096	4096	4096	4096	4096
XII	32	Two-bit	PAM4	16QAM	4096	4096	4096	4096	4096

Table 5.
Experimental design of BER tolerance evaluation.

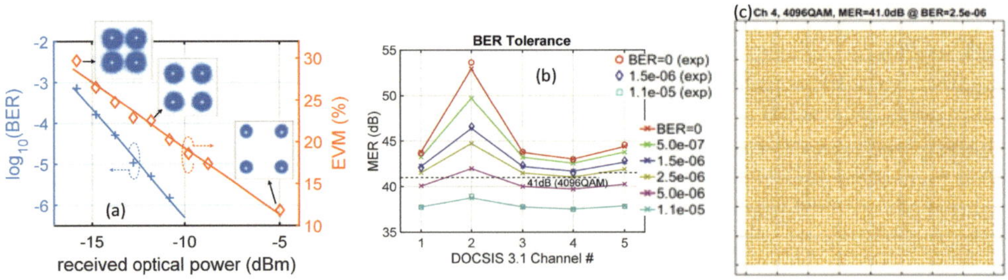

Figure 9.
BER tolerance of 32 GSa/s one-bit delta-sigma digitization: (a) BER and EVM versus received optical power; (b) MER degradation as BER increases; (c) 4096QAM constellation at BER threshold.

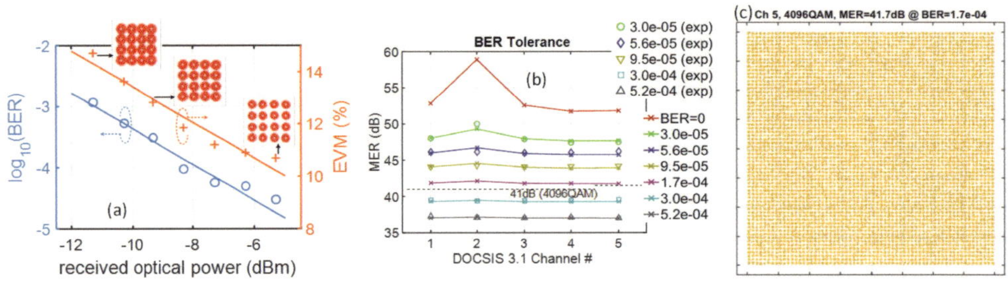

Figure 10.
BER tolerance of 32 GSa/s two-bit delta-sigma digitization: (a) BER and EVM versus received optical power; (b) MER degradation as BER increases; (c) 4096QAM constellation at the BER threshold.

of BER, 0, 1.5×10^{-6}, and 1.1×10^{-5} were tested, and good consistency is achieved between simulation and experimental results. As BER increases to 2.5×10^{-6}, MER of Channel 4 (worst case) drops to 41 dB, i.e., the minimum CNR requirement of 4096QAM. Therefore, the BER threshold of 32 GSa/s one-bit digitization is 2.5×10^{-6}. For BER < 2.5×10^{-6}, all 20 channels have sufficient CNR to support 4096QAM, but for BERs beyond this threshold, some channels' performance will drop below the DOCSIS 3.1 criteria. **Figure 9(c)** shows the 4096QAM constellation at the BER threshold.

Similar results for two-bit delta-sigma digitization (Case XII) are shown in **Figure 10**. The experimentally measured EVM and BER of 32 Gbaud DP-16QAM are shown in **Figure 10(a)**, and the minimum achievable BER is 3×10^{-5}. **Figure 10(b)** shows the MER performance degradation as a function of increasing BER. Simulation results (solid lines) are obtained by flipping a certain number of bits after digitization to emulate various BER values; experimental results (dots) at BER = 3×10^{-5}, 5.6×10^{-5}, 9.5×10^{-5}, 3×10^{-4}, 5.2×10^{-4} are obtained by adjusting the received optical power. Thanks to the additional quantization bits, two-bit digitization has larger tolerance against BER. Its MER performance becomes more resilient against the increasing BER, and the BER threshold is increased to 1.7×10^{-4}. The 4096QAM constellation at the threshold BER is shown in **Figure 10(c)**.

6. Conclusions

In this paper, we proposed and demonstrated delta-sigma digitization and coherent transmission of DOCSIS 3.1 signals for the first time. A low-pass CRFB delta-sigma ADC was designed to digitize 20 × 192-MHz DOCSIS 3.1 channels with 16384QAM and transmit the digitized RF spectrum over 80 km of fiber via a single-λ 128-Gb/s DP-QPSK or 256-Gb/s DP-16QAM coherent fiber link. Both one-bit and two-bit digitization were realized and supported by QPSK and 16QAM coherent links, respectively. To facilitate its application in HFC networks, a low-cost coherent fiber system was built based on narrowband optical modulator and RF amplifiers. Frequency pre-equalization was exploited to compensate the bandwidth limitation. MER larger than 50 dB was achieved for all 20 DOCSIS 3.1 channels, and high order modulation up to 16384QAM was supported. To date, this is the highest order of modulation transmitted over fiber in HFC networks. BER tolerance of delta-sigma digitization was also evaluated and negligible degradation of MER performance was observed for BER values up to 1.5×10^{-6} and 1.7×10^{-4}, for one-bit and two-bit digitization, respectively.

Acronyms

BBU	baseband unit
BDF/BDR	baseband digital forward/return
BER	bit error ratio
BS	base station
CCAP	converged cable access platform
CD	chromatic dispersion
CFO	carrier frequency offset
CM	cable modem
CMA	constant modulus algorithm
CMMA	constant multiple modulus algorithm
CNR	carrier-to-noise ratio
CPR	carrier phase recovery
CPRI	common public radio interface
C-RAN	cloud-radio access networks
CRFB	cascade resonator feedback
DOCSIS	data over cable system interface specification
DP	dual-polarization
DSL	digital subscriber line
DSP	digital signal processing
EPON	Ethernet passive optical network
E-UTRA	evolved universal terrestrial radio access
EVM	error vector magnitude
FTTH	fiber-to-the-home
GPON	gigabit passive optical network
HFC	hybrid fiber coax
HPF	high-pass filter
I	in-phase
IM/DD	intensity modulation/direct detection
MBH/MFH	mobile backhaul/fronthaul
MER	modulation error ratio
ML	maximum likelihood
MME	mobile management entity
MZM	Mach-Zehnder modulator
NG-PON2	next-generation passive optical network stage 2
NTF	noise transfer function
OFDM	orthogonal frequency division multiplexing

OOK	on-off keying
PAM4	four-level pulse-amplitude-modulation
PAPR	peak-to-average power ratio
PON	passive optical networks
Q	quadrature
QAM	quadrature amplitude modulation
QPSK	quadrature phase shift keying
RoF	radio-over-fiber
RPD	remote PHY device
RRH	remote radio heads
SC-QAM	single-carrier quadrature amplitude modulation
S-GW	service gateway
SNR	signal-to-noise ratio
TDM	time-division-multiplexing
TWDM	time- and wavelength-division multiplexing
TWDM-PON	time- and wavelength-division multiplexed passive optical network

Author details

Jing Wang*, Zhensheng Jia, Luis Alberto Campos and Curtis Knittle
CableLabs, Louisville, CO, United States

*Address all correspondence to: j.wang@cablelabs.com

References

[1] Kramer G, Pesavento G. Ethernet passive optical network (EPON): Building a next-generation optical access network. IEEE Communications Magazine. 2002;**40**(2):66-73

[2] Park S-J, Lee C-H, Jeong K-T, Park H-J, Ahn J-G, Song K-H. Fiber-to-the-home services based on wavelength-division-multiplexing passive optical network. Journal of Lightwave Technology. 2004;**22**(11):2582-2591

[3] Luo Y, Zhou X, Effenberger F, Yan X, Peng G, Qian Y, et al. Time- and wavelength-division multiplexed passive optical network (TWDM-PON) for next-generation PON stage 2 (NG-PON2). Journal of Lightwave Technology. 2013;**31**(4):587-593

[4] China Mobile. C-RAN the Road towards Green RAN (Version 2.5). White Paper; Oct 2011

[5] Checko A, Christiansen HL, Yan Y, Scolari L, Kardaras G, Berger MS, et al. Cloud RAN for mobile networks—A technology overview. IEEE Communications Surveys & Tutorials. 2015;**17**(1):405-426

[6] Pfeiffer T. Next generation mobile fronthaul and midhaul architectures. IEEE/OSA Journal of Optical Communications and Networking. 2015;**7**(11):B38-B45

[7] Pizzinat A, Chanclou P, Saliou F, Diallo T. Things you should know about fronthaul. IEEE Journal of Lightwave Technology. 2015;**33**(5):1077-1083

[8] Toy M. Cable Networks, Services, and Management. IEEE Press Series on Networks and Services Management. John Wiley & Sons; 2015. ISBN-10: 1118837592, chapter 1

[9] Ovum report. HFC: Delivering Gigabit Broadband. http://www.nbnco.com.au/content/dam/nbnco2/documents/HFC%20Delivering%20Gigabit%20Broadband%20Ovum%20Report.pdf

[10] Hamzeh B, Toy M, Fu Y, Martin J. DOCSIS 3.1: Scaling broadband cable to gigabit speeds. IEEE Communications Magazine. 2015;**53**(3):108-113

[11] Toy M, Martin J, Schmitt M, Blake V. Next generation cable networks with DOCSIS 3.1 technology. IEEE Communications Magazine. 2015;**53**(3):106-107

[12] Mehmood H, Rahman S, Cioffi JM. Bit loading profiles for high-speed data in DOCSIS 3.1. IEEE Communications Magazine. 2015;**53**(3):114-120

[13] Rice D. DOCSIS 3.1 technology and hybrid fiber coax for multi-Gbps broadband. In: Optical Fiber Communication Conference (OFC) 2015; paper Th4B.1

[14] Salib H. FTTU: MSO perspective. Optical Fiber Communications Conference (OFC) 2016; paper Th1I.1

[15] Cable Television Laboratories, Inc. Data-Over-Cable Service Interface Specifications, DOCSIS 3.1. Physical Layer Specification, CM-SP-PHYv3.1-I09-160602; Jun 2016

[16] Cable Television Laboratories, Inc. DOCSIS 3.1 Physical & MAC Layer Quick Reference Pocket Guide; Sep 2014

[17] Sundaresan K. DOCSIS 3.1 High Level Overview at NANOG 59. Cable Television Laboratories, Inc; 2013. https://www.nanog.org/sites/default/files/wed.general.sundaresan.docsis.35.pdf

[18] Al-Banna A, Cloonan T. The spectral efficiency of DOCSIS 3.1 systems. Arris

White Paper; 2014. https://www.arris.com/globalassets/resources/white-papers/arris_spectral_efficiency_of_docsis_wp.pdf

[19] Ibl H, Klaus C. DOCSIS 3.1 application note. Rohde Schwarz White Paper, 7MH89_0E; 2015. http://www.rohde-schwarz-usa.com/rs/rohdeschwarz/images/7mh89_oe-docsis3. 1.pdf

[20] 3GPP TR 36.808, V10.1.0. Evolved Universal Terrestrial Radio Access (E-UTRA); Carrier Aggregation; Base Station (BS) radio transmission and reception. Release 10, 2013-07

[21] 3GPP RP-142286. New WI proposal: LTE Carrier Aggregation Enhancement Beyond 5 Carriers. 2014-12

[22] Wang J, Liu C, Zhu M, Yi A, Cheng L, Chang GK. Investigation of data-dependent channel cross-modulation in multiband radio-over-fiber systems. IEEE Journal of Lightwave Technology. 2014;**32**(10):1861-1871

[23] Wang J, Liu C, Zhu M, Xu M, Dong Z, Chang G-K. Characterization and mitigation of nonlinear intermodulations in multichannel OFDM radio-over-fiber systems. In: European Conference on Optical Communication (ECOC) 2014, paper P.7.18

[24] Wang J, Liu C, Zhang J, Zhu M, Xu M, Lu F, et al. Nonlinear inter-band subcarrier intermodulations for multi-RAT OFDM wireless services in 5G heterogeneous mobile fronthaul networks. IEEE Journal of Lightwave Technology. 2016;**34**(17):4089-4103

[25] Common Public Radio Interface (CPRI). Specification V7.0 (2015-10-09). http://www.cpri.info/downloads/CPRI_v_7_0_2015-10-09.pdf

[26] Gray R. Oversampled sigma-delta modulation. IEEE Transactions on Communications. 1987;**35**(5):481-489

[27] Aziz PM, Sorensen HV, der Spiegel JV. An overview of sigma-delta converters. IEEE Signal Processing Magazine. 1996;**13**(1):61-84

[28] Schreier R, Temes GC. Understanding Delta-Sigma Data Converters. Wiley-IEEE Press; Nov. 2004. 464 pp. ISBN: 978-0-471-46585-0

[29] Schreier R. High-level design and simulation of delta-sigma modulators. http://www.mathworks.com/matlabcentral/fileexchange/19-delta-sigma-toolbox

[30] Hung TP, Rode J, Larson LE, Asbeck PM. Design of H-bridge class-D power amplifiers for digital pulse modulation transmitters. IEEE Transactions on Microwave Theory and Techniques. 2007;**55**(12):2845-2855

[31] Nielsen M, Larsen T. A transmitter architecture based on delta-sigma modulation and switch-mode power amplification. IEEE Transactions on Circuits and Systems II: Express Briefs. 2007;**54**(8):735-739

[32] Ghannouchi FM, Hatami S, Aflaki P, Helaoui M, Negra R. Accurate power efficiency estimation of GHz wireless delta-sigma transmitters for different classes of switching mode power amplifiers. IEEE Transactions on Microwave Theory and Techniques. 2010;**58**(11):2812-2819

[33] Jerng A, Sodini CG. A wideband ΔΣ digital-RF modulator for high data rate transmitters. IEEE Journal of Solid-State Circuits. 2007;**42**(8):1710-1722

[34] Frappe A, Flament A, Stefanelli B, Kaiser A, Cathelin A. An all-digital RF signal generator using high-speed ΔΣ modulators. IEEE Journal of Solid-State Circuits. 2009;**44**(10):2722-2732

[35] Pozsgay A, Zounes T, Hossain R, Boulemnakher M, Knopik V, Grange S. A fully digital 65 nm CMOS transmitter for the 2.4-to-2.7GHz WiFi/WiMAX bands using 5.4 GHz ΔΣ RF DACs. In: IEEE International Solid-State Circuits Conference (ISSCC). 2008. pp. 360-361

[36] Silva NV, Oliveira ASR, Carvalho NB. Design and optimization of flexible and coding efficient all-digital RF transmitters. IEEE Transactions on Microwave Theory and Techniques. 2013;**61**(1):625-632

[37] Chung S, Ma R, Shinjo S, Nakamizo H, Parsons K, Teo KH. Concurrent multiband digital outphasing transmitter architecture using multidimensional power coding. IEEE Transactions on Microwave Theory and Techniques. 2015;**63**(2):598-613

[38] Wepman JA. Analog-to-digital converters and their applications in radio receivers. IEEE Communications Magazine. May 1995;**33**(5):39-45

[39] Miller MR, Petrie CS. A multibit sigma-delta ADC for multimode receivers. IEEE Journal of Solid-State Circuits. 2003;**38**(3):475-482

[40] Arias J, Kiss P, Prodanov V, Boccuzzi V, Banu M, Bisbal D, et al. A 32-mW 320-MHz continuous-time complex delta-sigma ADC for multi-mode wireless-LAN receivers. IEEE Journal of Solid-State Circuits. 2006;**41**(2):339-351

[41] Wu C, Alon E, Nikolić B. A wideband 400 MHz-to-4 GHz direct RF-to-digital multimode ΔΣ receiver. IEEE Journal of Solid-State Circuits. 2014;**49**(7):1639-1652

[42] Bettini L, Christen T, Burger T, Huang Q. A reconfigurable DT ΔΣ modulator for multi-standard 2G/3G/4G wireless receivers. IEEE Journal on Emerging and Selected Topics in Circuits and Systems. 2015;**5**(4):525-536

[43] Kitamura K, Sasaki S, Matsuya Y, Douseki T. Optical wireless digital-sound transmission system with 1-bit ΔΣ-modulated visible light and spherical Si solar cells. IEEE Sensors Journal. 2010;**10**(11):1753-1758

[44] Qian H, Chen J, Yao S, Yu Z, Zhang H, Xu W. One-bit sigma-delta modulator for nonlinear visible light communication systems. IEEE Photonics Technology Letters. Feb. 2015;**27**(4):419-422

[45] Kanno A, Kawanishi T. Analog signal transmission by FPGA-based pseudo-delta-sigma modulator. In: IEEE Photonics Conference (IPC). 2015. pp. 136-137

[46] Jang S, Jo G, Jung J, Park B, Hong S. A digitized IF-over-fiber transmission based on low-pass delta-sigma modulation. Photonics Technology Letters. 2014;**26**(24):2484-2487

[47] Pessoa LM, Tavares JS, Coelho D, Salgado HM. Experimental evaluation of a digitized fiber-wireless system employing sigma delta modulation. Optics Express. 2014;**22**(14):17508-17523

[48] Wang J, Yu Z, Ying K, Zhang J, Lu F, Xu M, et al. Delta-sigma modulation for digital mobile fronthaul enabling carrier aggregation of 32 4G-LTE/30 5G-FBMC signals in a single-λ 10-Gb/s IM-DD channel. In: Optical Fiber Communication Conference (OFC). 2016. paper W1H.2

[49] Wang J, Yu Z, Ying K, Zhang J, Lu F, Xu M, et al. 10-Gbaud OOK/PAM4 digital mobile fronthaul based on one-bit/two-bit delta-sigma modulation supporting carrier aggregation of 32 LTE-A signals. European Conference on Optical Communication (ECOC). 2016. paper W.4.P1.SC7.1

[50] Wang J, Yu Z, Ying K, Zhang J, Lu F, Xu M, et al. Digital mobile fronthaul

based on delta-sigma modulation for 32 LTE carrier aggregation and FBMC signals. IEEE/OSA Journal of Optical Communications and Networking. 2017;**9**(2):A233-A244

[51] Wang J, Jia Z, Campos LA, Knittle C, Chang G. Optical coherent transmission of 20x192-MHz DOCSIS 3.1 channels with 16384QAM based on delta-sigma digitization. In: Optical Fiber Communication Conference (OFC). 2017. paper Th1K.1

[52] Cable Television Laboratories, Inc. Data-over-cable service interface specifications, DCA-MHAv2, Remote PHY Specification. CM-SP-R-PHY-I07-170524; May 2017

[53] Salinger JD. Remote PHY: Why and how. CableLabs, NCTA, SCTE, Spring Technical Forum; 2014

[54] Chapman JT. Remote PHY for Converged DOCSIS, Video and OOB. CableLabs, NCTA, SCTE, Spring Technical Forum; 2014

[55] Chapman JT, White G, Jin H. Impact of CCAP to CM Distance in a Remote PHY Architecture. CableLabs, NCTA, SCTE, Spring Technical Forum; 2015

[56] Salinger JD. Distributed Architectures and Converged Access Network. CableLabs, NCTA, SCTE, Spring Technical Forum; 2016

[57] Chih-Lin I, Huang J, Yuan Y, Ma S, Duan R. NGFI, the xHaul. In: IEEE Globecom 2015 Workshops

[58] Chih-Lin I, Yuan Y, Huang J, Ma S, Cui C, Duan R. Rethink fronthaul for soft RAN. IEEE Communications Magazine. 2015;**53**(9):82-88

[59] Chih-Lin I, Huang J. RAN revolution with NGFI (xHaul) for 5G. In: Optical Fiber Communications Conference (OFC). p. 2017. paper W1C.7

[60] Haykin SS. Adaptive Filter Theory. Englewood Cliffs, NJ: Prentice-Hall; 2001

[61] Savory SJ. Digital filters for coherent optical receivers. Optics Express. 2008;**16**(2):804-817

[62] Savory SJ. Digital coherent optical receivers: Algorithms and subsystems. IEEE Journal of Selected Topics in Quantum Electronics. 2010;**16**(5):1164-1179

[63] Godard D. Self-recovering equalization and carrier tracking in two-dimensional data communication systems. IEEE Transactions on Communications. 1980;**28**(11):1867-1875

[64] Kikuchi K. Polarization-demultiplexing algorithm in the digital coherent receiver. In: Digest of IEEE/LEOS Summer Topical Meetings. 2008. pp. 101-102

[65] Leven A, Kaneda N, Koc UV, Chen YK. Frequency estimation in intradyne reception. IEEE Photonics Technology Letters. 2007;**19**(6):366-368

[66] Hoffmann S, Bhandare S, Pfau T, Adamczyk O, Wordehoff C, Peveling R, et al. Frequency and phase estimation for coherent QPSK transmission with unlocked DFB lasers. IEEE Photonics Technology Letters. 2008;**20**(18):1569-1571

[67] Viterbi AJ, Viterbi AM. Nonlinear estimation of PSK-modulated carrier phase with application to burst digital transmission. IEEE Transactions on Information Theory. July 1983;**29**(4):543-551

[68] Fatadin I, Ives D, Savory SJ. Blind equalization and carrier phase recovery in a 16-QAM optical coherent system. IEEE Journal of Lightwave Technology. 2009;**27**(15):3042-3049